Religions and
Developm

Religion has been excluded from development studies for decades. Religious traditions have contributed greatly towards development work, yet major international players have tended to ignore its role. However, recent years have shown a noticeable shift in development policy, practice and research to recognize religion as a relevant factor.

This text provides a comprehensive insight into different approaches towards understanding the relationships between religions and development studies, policy and practice. It guides readers through current debates, presenting, explaining and critically evaluating a broad range of literature and locating it within a theoretical context. The text explores the role of religion within development, from positive contributions, such as the important role that many 'faith-based organizations' play in education or health care, to more complicated and contested notions of impact, such as religiously inspired violence or gender inequality. The book begins with three background chapters that outline the relevance of religions for development studies, policy and practice, and introduce the reader to the study of 'development' and of 'religions'. Following these, the focus then shifts to examine a number of thematic areas, including religion, gender and development, and the implications of the 'rise of religion' for mainstream development studies, policy and practice in the twenty-first century. Each chapter contains a range of features to assist undergraduate learning, including learning objectives for each chapter, discussion of key concepts, summaries, discussion questions, further reading and websites. The book also contains over sixty boxed case studies to provide further definition, explanation, and examples of the interactions between religions and development globally.

This innovative text presents religions as something that can both obstruct and aid development, encouraging readers to engage critically with the multiple ways that religion impacts on both the conceptualization of development and the resulting project interventions. It will be of interest to undergraduate and postgraduate students and scholars interested in religious studies, development studies and the broader study of societies and cultures.

Emma Tomalin is Senior Lecturer in Religious Studies at the School of Philosophy, Religion and the History of Science at the University of Leeds. Her research interests are focused around religion and global development, and religion, gender and society.

Routledge Perspectives on Development

Series Editor: Professor Tony Binns, *University of Otago*

Since it was established, in 2000, the same year as the Millennium Development Goals were set by the United Nations, the *Routledge Perspectives on Development* series has become the pre-eminent international textbook series on key development issues. Written by leading authors in their fields, the books have been popular with academics and students working in disciplines such as anthropology, economics, geography, international relations, politics and sociology. The series has also proved to be of particular interest to those working in interdisciplinary fields, such as area studies (African, Asian and Latin American studies), development studies, environmental studies, peace and conflict studies, rural and urban studies, travel and tourism.

If you would like to submit a book proposal for the series, please contact the Series Editor, Tony Binns, on: jab@geography.otago.ac.nz

Published:

Third World Cities, 2nd Edition
David W. Drakakis-Smith

Rural–Urban Interactions in the Developing World
Kenneth Lynch

Environmental Management and Development
Chris Barrow

Tourism and Development
Richard Sharpley and David J. Telfer

Southeast Asian Development
Andrew McGregor

Population and Development
W. T. S. Gould

Postcolonialism and Development
Cheryl McEwan

Conflict and Development
Andrew Williams and Roger MacGinty

Disaster and Development
Andrew Collins

Non-Governmental Organisations and Development
David Lewis and Nazneen Kanji

Cities and Development
Jo Beall

Gender and Development, 2nd Edition
Janet Momsen

Economics and Development Studies
Michael Tribe, Frederick Nixson and Andrew Sumner

Water Resources and Development
Clive Agnew and Philip Woodhouse

Theories and Practices of Development, 2nd Edition
Katie Willis

Food and Development
E. M. Young

An Introduction to Sustainable Development, 4th Edition
Jennifer Elliott

Latin American Development
Julie Cupples

Religions and Development
Emma Tomalin

Development Organizations
Rebecca Shaaf

Forthcoming:

Global Finance and Development
David Hudson

Natural Resource Extraction and Development
Roy Maconachie and Gavin M. Hilson

Children, Youth and Development, 2nd Edition
Nicola Ansell

Climate Change and Development
Thomas Tanner and Leo Horn-Phathanothai

Politics and Development
Heather Marquette and Tom Hewitt

South Asian Development
Trevor Birkenholtz

Conservation and Development
Shonil Bhagwat and Andrew Newsham

Religions and Development

Emma Tomalin

Routledge
Taylor & Francis Group

LONDON AND NEW YORK

First published 2013
by Routledge
2 Park Square, Milton Park, Abingdon, Oxon OX14 4RN

Simultaneously published in the USA and Canada
by Routledge
711 Third Avenue, New York, NY 10017

Routledge is an imprint of the Taylor & Francis Group, an informa business

British Library Cataloguing in Publication Data
A catalogue record for this book is available from the British Library

Library of Congress Cataloging-in-Publication Data
Tomalin, Emma.
Religions and development / Emma Tomalin.
pages cm
Includes bibliographical references and index.
1. Developing countries–Religion. 2. Religions. 3. Religion. I. Title.
BL2680.T66 2013
201'.73091724–dc23
2012041590

ISBN: 978-0-415-61349-1 (hbk)
ISBN: 978-0-415-61350-7 (pbk)
ISBN: 978-0-203-83117-5 (ebk)

Typeset in Times New Roman
by Keystroke, Station Road, Codsall, Wolverhampton

Printed and bound in Great Britain by
TJ International Ltd, Padstow, Cornwall

Contents

Plates and figures

Plates

Figures

Tables

Boxes

Acknowledgements

My ability to write this book was largely the product of my involvement in the DFID-funded Religions and Development (RaD) research programme, based in the International Development Department at Birmingham University between 2005 and 2010. The research that we undertook during this programme made a significant contribution to this important area within development studies and practice, and played a key role in charting and analysing the 'turn to religion' by development actors discussed in this volume. During this programme I attended many meetings with colleagues from across the United Kingdom and also internationally, and the conversations that took place have contributed towards the shaping of my knowledge about and approach to the study of 'religions and development'. I would like to thank the anonymous reviewer of the manuscript for constructive and encouraging comments, and also Katherine Marshall (Visiting Professor, School of Foreign Service, Georgetown University; Executive Director, World Faiths Development Dialogue) for her extremely helpful feedback. I would like to extend particular thanks to Professor Carole Rakodi (also the director of the RaD programme), who reviewed the manuscript and continued to provide support and comments as I undertook the final revisions of the book. Faye Leerink, the senior editorial assistant for geography at Routledge, has provided prompt and supportive feedback throughout the process, and I would like to specifically thank her for assisting me with selecting the photographs included in the book. My undergraduate and Master's students in Theology and Religious Studies at Leeds, enrolled on my 'Religions and Global Development' modules, have patiently read draft chapters of the volume, at times offering useful comments from the student point of view. Finally, I would like to mention my family – Andy, Reuben, Toby and Theo – for frequently providing a much-needed distraction from the job at hand, while also allowing me to hide away over much of the summer months to complete the final revisions to the text.

● 1 Introduction: religions and development – a new agenda?

This chapter will:

- outline the relevance of the study of 'religions and development';
- analyse the reasons why religion has been ignored or marginalized by mainstream development studies, policy and practice;
- discuss the reasons for the recent 'turn to religion' by international development researchers and actors;
- provide an overview of the book's structure, approach and content.

Introduction: the relevance of the study of 'religions and development'

> For most people of the 'South' spirituality is integral to their understanding of the world and their place in it, and so is central to the decisions that they make about their own and their communities' development. Their spirituality affects decisions about who should treat their sick child, when and how they will plant their fields, and whether or not to participate in risky but potentially beneficial social action. Despite the evident centrality of spirituality to such decisions, the subject is conspicuously under-represented in the development discourse.
>
> (Ver Beek 2000: 31)

While religion is an important dimension of the lives of many people across the globe, until recently it has tended to be ignored or marginalized by international development research, policy and practice. Whether we are considering 'religion', 'spirituality' or 'faith', the donor-driven development project that emerged in the Global North after the Second World War has typically viewed these aspects of life as outdated and likely to disappear as societies develop and modernize. Additionally, 'culture' more broadly, of which religion is often viewed as a manifestation, has been understood as something that potentially gets in the way of development (see Box 1.1 for a discussion of the relationship between religion and culture). As Eade writes, '"local" or "traditional" cultures are even now seen as a brake on development, while the international development agencies and their national counterparts regard themselves as culturally neutral – if not superior' (2002: ix). Moreover, according to Eade, development policies and practices 'for the most part . . . proceed as though all cultures are, or seek to be, more or less the same: development, from this perspective, is a normative project' (ibid.).

Box 1.1

The relationship between 'religion' and 'culture'

There is a lack of consensus about the relationship between religion and culture, and the degree to which it is possible to distinguish between them. Anthropologists, who typically focus on culture in their work, tend to view religion as part of a wider nexus of practices and attitudes that make up 'culture'. To use Tylor's definition, which is one among many contested and often competing definitions: 'culture, or civilization, taken in its broad, ethnographic sense, is that complex whole which includes knowledge, belief, art, morals, law, custom, and any other capabilities and habits acquired by man [*sic*] as a member of society' (1958 [1871]: 1).

Thus, 'culture' is a term used to denote products and activities (art, music, etc.) as well as a way of life more broadly. It is also used in the sense of referring to 'a culture': a distinct group of people united by shared ways of seeing the world. More recently, anthropologists have criticized older understandings of 'culture' for implying that 'cultures' (as in groups of people) and 'culture' (as in artistic products and ways of life) are rather more static and bounded than they actually are. Such views of culture(s) are a legacy of colonial ways of seeing the world, which essentialize the non-West as timeless and traditional rather than evolving, progressive and modern. Instead, contemporary postcolonial or postmodern views of culture emphasize its contested and dynamic nature.

Definitions of religion have also been contested, with a broad distinction being drawn between *substantive* (what religion is) and *functionalist* (what religion does) definitions (these will be discussed in more detail in Chapter 3). The view that religion is a subset of culture, and that each influences the other, not only is found within anthropology but is popular within the social sciences more broadly. However, members of religious traditions may view the relationship between culture and religion somewhat differently, exhibiting a reluctance to reduce religion to culture. Instead of completely subsuming religion within a broader category of culture, religious practitioners often draw a distinction between the two to highlight practices and beliefs that they consider are not inherent to the religion, yet are typically carried out by members of the tradition. An example of this is the argument that female genital mutilation (FGM) is a 'cultural' practice followed by many Muslims rather than an Islamic 'religious' practice. This distinction is often made with respect to things that members of religious traditions wish to distance themselves from, particularly if they are discriminatory or harmful.

Moreover, members of religious traditions may not wish to view their religion as 'cultural', as this implies that it has been influenced and shaped by cultural processes, rather than being the timeless word of God, or reflecting the teachings of the original founder of the religious tradition. Others might accept that some aspects of their religious tradition have been influenced and shaped by surrounding cultural practices and attitudes. This perspective is popular within feminist theology, where there is a need to differentiate parts of the tradition that have been influenced by patriarchy from those that are beyond culture and reflect the 'true religion'.

However, as the opening quotation from Ver Beek suggests, in terms of day-to-day existence, 'spirituality' may be relevant to the decisions people make, and can provide meaningful answers to the questions that life poses, as well as comfort and support in times of difficulty. Religion also shapes social values and structures, such as gender and family dynamics, and influences political systems. Finally, religious institutions and organizations, from churches, mosques and temples to more recently formed 'faith-based organizations' (FBOs) (e.g. Islamic Relief or Christian Aid), are involved in various forms of charitable work, are often at the forefront of disaster relief and increasingly take part in 'development' activities that aim to improve people's lives in the long term. Thus, according to Ver Beek (2000) and others, an understanding of the roles that 'religion', 'faith' or 'spirituality' play in people's lives, shaping their beliefs and values, and the decisions that they make, as well as the impact of religion on political and social systems, should inform the work undertaken by development actors. Box 1.2 outlines some of the main reasons why it is important that international development actors take religions seriously.

Box 1.3 clarifies and defines the terms 'spirituality' and 'faith', which are often used interchangeably with the word 'religion'. Although the use of either of these terms instead of religion may be arbitrary, in some cases they are chosen because of their specific meaning or connotations. Reflecting upon why and when they are used, as well as who uses them, may tell us something about the perspective of the particular commentator. It may also be revealing to reflect upon the circumstances in which they are not used, and 'religion' is chosen instead. Overall, in this book the word 'religion' is predominantly used, rather than 'faith' or 'spirituality', unless referring to other people's use of the different terms. While the term and concept 'religion' is not uncontroversial (see Chapter 3), the term 'faith' is problematic for having stronger associations with Christian

---

Box 1.2

Why does it matter if international development actors ignore religion?

- The fullest range of factors blocking as well as facilitating 'development' may not be taken into consideration.
- Programmes and interventions could be imposed that will not have the desired outcomes or may be inappropriate for the particular community or context.
- There could be an increased perception of 'development' as a western, colonialist imposition.
- Development actors and agencies could fail to find out what sort of 'development' is needed and desired in particular contexts, since religion often helps shape people's views about this. This may challenge understandings of what counts as 'development' in the Global North.
- Valuable opportunities to work with religious actors and institutions, including FBOs, in the pursuit of development goals, could be missed.

Box 1.3

'Spirituality' and 'faith'

'Spirituality' is often understood to be part of what it is to be religious. A number of commentators define spirituality as that part of religion that is not institutionalized or focused on doctrine, but which captures the experience of the relationship between the individual and the sacred or the divine (King 1996; Zinnbauer *et al.* 1997). People often choose this term in preference to 'religion' when they wish to emphasize these aspects. For example, they may wish to draw attention to dimensions of religions that they consider to be universal or benign, in contrast to those aspects of religion that are dogmatic, may give rise to sectarianism or are illiberal.

The term 'faith' similarly attempts to capture the dimension of religions that extends beyond the codification of values, rules and social practices within particular traditions. As with the term 'spirituality', there is a tendency in some parts of the world for it to be increasingly used as a substitute for the word 'religion', owing to the apparent growth in numbers of people (particularly in the Global North) who still have belief or faith in a divine being or a personal spirituality, but outside the confines of institutionalized religion. The widespread use of the term can be traced to the United States specifically, where in the latter decades of the twentieth century 'faith' became a more significant part of public discourse and policy agendas (e.g. through the White House Office of Faith-Based and Community Initiatives (OFBCI), established by President George W. Bush in 2001).

The term 'faith' has come to have a very significant currency within development discourse and frequently serves as a substitute for 'religion' (for example, the term 'faith-based organizations' is the most common word for religiously inspired organizations that work in development). It tends to be the preferred term for organizations and agencies that are attempting to work with different religions, such as the 'World Faiths Development Dialogue' and USAID's 'Faith-Based and Community Initiatives' (Box 8.5 on p. 214). The extent to which this term has the same relevance outside western Christian contexts has been questioned. Jeavons, for instance, suggests that 'faith' in these debates relates 'more closely with the religious tradition(s) of Christianity and distances [them] from others – like Islam or Hinduism – in which, in fact, faith is not a particularly meaningful concept or term' (2004: 141). Kirmani and Zaidi (2010), in their study of Muslim charities in Karachi, found that 'faith-based' was not a term that was used by these charities to describe their work (Chapter 8).

---

traditions than with other religions. Moreover, as Box 1.3 shows, the term 'spirituality' is often selected to refer to a dimension of religion that is not institutionalized or focused on doctrine. However, religious institutions and doctrines are precisely of interest within this book, not least since they can shape conceptions of what development should consist of, which may or may not cohere with the visions of development pursued by development actors in the Global North.

One aim of this book is to provide an understanding of the reasons for the uneasy relationship between religion and development (Selinger 2004). But what do we mean by 'development' here? The book primarily focuses on mainstream development studies, policy and practice, driven by governments, international agencies, academics and activists, mainly in the Global North, whose initial motivation was a sense of responsibility towards ex-colonies (see Chapter 2 for a more detailed discussion of different meanings of and approaches to development). Religion is also relevant to thinking about other meanings of development, which need to become part of the analysis but are not the main focus here. These include understandings of development as referring to broad processes of social change, rather than the more specific donor-led development projects that emerged after the Second World War. It is also important to look at development from the perspective of developing country governments and organizations, which may interact with the global aid business to a greater or lesser degree. Although the topic is an important one, this book does not focus on the role of religion in national development processes and programmes. While we should not collapse 'development' into the conception of it advanced by the 'global aid business', it is important to undertake a critical analysis of this business, particularly with respect to its failure to engage (at least until recently) with cultural and religious diversity and the actual and potential links between culture and religion and the goals and practices of development. Thus, this book takes as its primary audience those researching and working in development in (predominantly) Global North contexts, whose inherited models or approaches to development have made it difficult to consider the relevance of religion to the achievement of development aims.

Secularization and religious resurgence

The early decades of western[1] development aid focused on economic growth and it was assumed that religious worldviews were inimical to economic development and that as societies modernized, they would become less religious. However, as Table 1.1 suggests, rather than disappearing or diminishing in significance, religion continues to exist alongside modernizing and globalizing processes, often adapting and even intensifying in response to changing social, economic and political environments. Even in so-called developed nations, religious belief and practice is an important aspect of many people's lives. Chapter 3 will look at critiques of such simplistic theories of secularization, which argue that the increased, renewed or continuing role of religion in public life, as well as in people's private lives, means that the relationships between religions and society are both complex and controversial, and should be taken seriously.

Debates about the 'global resurgence of religion' have been building momentum since the 1980s, if not earlier (Thomas 2005), including the extent to which there has been an actual resurgence or just the appearance of one because western governments and the media are taking more notice of religions globally. The Iranian Islamic Revolution of 1979 was in many ways a wake-up call to

Table 1.1 *The world's fastest-growing religions*

Religion	Growth rate over the period from 2000 to 2005	Adherents worldwide	Behind the trend
Islam	1.84%	1.3 billion	High birth rates in Asia, the Middle East and Europe
Baha'i	1.7%	7.7 million	High birth rates in India
Sikhism	1.62%	25.8 million	High birth rates in India
Jainism	1.57%	5.9 million	High birth rates in India
Hinduism	1.52%	870 million	High birth rates in India
Christianity	1.38%	2.2 billion	High birth rates and conversions in the Global South (particularly, Pentecostal movements in Latin America, Africa, China, and India).

Source: Foreign Policy (2007)

Plate 1.1 *A large group of women protestors in Tehran, Iran, holding photos of the religious leader of the 1979 Iranian Revolution, Ayatollah Ruhollah Musavi Khomeini, November 1979*

Source: Reza/Webistan.

Protesters participate at a demonstration in support of the US hostage crisis. From 4 November 1979 to 20 January 1981, for 444 days, 52 US diplomats were held hostage after a group of Islamist students took over the American embassy in support of the Iranian Revolution.

those who considered that religion had lost its public influence (Plate 1.1). Far from being an exception, the following decades have witnessed many examples of religiously inspired social and political action across the globe – from the *Satanic Verses* controversy in the early 1990s to the influence of the Catholic Church on reproductive rights issues, the rise of nationalist Hindu politics in South Asia and the watershed event of 9/11 (Plates 1.2 and 1.3). A range of different factors have been suggested to account for this resurgence of religion globally, including the rise of the 'religious right' in the United States following the election of Ronald Reagan as US president in 1980, which began a process of patronage for evangelical Christianity that transformed US politics, including foreign policy (Clarke 2008: 19). As a counter-current to the US Christian right, the rise of political Islam is also significant, spurred by the Iranian Revolution and the support of the US government for the *mujahideen* ('Islamic holy warriors') in fighting the Soviet occupation of Afghanistan after 1979. According to Thomas, this Afghan war mobilized and radicalized 'an entire transnational generation of Muslim youths prepared to die for Kashmir, Palestine, Chechnya and the Taliban's Afghanistan' (2005: 3). Finally, the collapse of communism and the end of the Cold War played a role in the resurgence of religion, which, as Clarke suggests, 'fuelled the rise of identity politics, centred on novel blends

Plate 1.2 *Anti-Salman Rushdie demonstration in Hyde Park, London*

Source: Janine Wiedel Photolibrary/Alamy.

In 1988, Salman Rushdie's novel *The Satanic Verses* was published in the United Kingdom. The book was perceived by many Muslims to be blasphemous and in 1989 a *fatwa* was issued by the Iranian ayatollah Ruhollah Musavi Khomeini ordering Muslims to kill Rushdie.

Plate 1.3   *Bajrang Dal volunteers display 'trishuls'*

Source: AFP/Getty Images.

Bajrang Dal is the youth wing of the Vishva Hindu Parishad (VHP), a Hindu nationalist organization based on the ideology of *Hindutva* (see Box 4.8). In this image, Bajrang Dal volunteers are displaying 'trishuls', or tridents, which they received on the last day of a seven-day youth camp at the Koteshwar Swaminarayan Gurukul on the outskirts of Ahmedabad on 20 May 2012.

of ethnic, cultural and religious identity' (2008: 22) for those political activists previously attracted to socialism and communism.

These debates about secularization and the resurgence of religion have had relatively little impact upon the development community until recently. The assumption that religion would disappear or that it was, on the whole, problematic prevailed (for example, in family planning most international or bilateral agencies saw religion only as a problem). Over the past decade there has been a noticeable shift within development studies, policy and practice to reflect more fully the relevance of religion to development processes and outcomes, not only as a negative force but also as a potentially positive factor, and there is now an increasing call for development to be 'embedded in local culture' (Verhelst and Tyndale 2002: 11). As Haynes writes:

> [R]ecently, this negative consensus about religion has fractured, partly as a result of the widespread failure of secular development trajectories to achieve widespread poverty reduction or reductions of inequality and injustice in the developing world. As a result, the emphasis of development thinking has shifted to include various religious expressions that are now widely seen as potentially

important components of achieving development g
associated with 'religious extremism' are not.

The realization that there may be different, yet valid, approach   ..ι
that are influenced by a community's religion and culture are i ..casingly given
attention. Development studies journals have witnessed an upturn in the
publication of research articles that deal with the relationships between religions
and international development, from a situation where this topic was more or
less ignored (see Table 2.1). There has also been a shift in attitude among
western donors about including religious voices in development debates and
initiatives. For instance, the views of religious leaders have been invited on key
initiatives such as the Millennium Development Goals (Clarke and Jennings
2008a: 2). There has also been an increased engagement between western
donors and FBOs. Although many development agencies had been funding
FBOs for years, often without explicitly acknowledging their religious connections, more recently, as Rakodi writes, 'they started to pay attention to the role
of religion, including their own relationships with religious organizations'
(2011: 5). Moreover, in recent years there has been a growing commitment
to funding FBOs. The United States has gone furthest, with George W.
Bush, during his presidency, almost doubling funding to 'faith-based' groups,
from 10.5 per cent of aid in 2001 to 19.9 per cent in 2005 (James 2009).
Following this lead, in the United Kingdom the Department for International
Development's (DFID) 2009 White Paper promised to double funding to FBOs
(James 2011), reflecting 'recognition' of the 'unique contribution that they can
make in both delivering development on the ground, and connecting with communities here and abroad' (DFID 2009: 134). Box 1.4 outlines some examples
of this 'turn to religion' in development policy and practice in the Global North.

Box 1.4

The 'turn to religion' in development settings in the Global North: some examples

In the United States, as well as a number of European settings, the role of religion in
'development cooperation' is beginning to be taken more seriously. In the United States
the rise of the 'religious right' since the 1980s and reactions after 9/11 have both
contributed to a greater focus on the role of 'faith' in society, including increased funding
for FBOs. In 2001 the *Center for Faith-Based and Community Initiatives* (CFBCI) was
established at USAID 'to create a level playing field for faith and community based
organizations to compete for USAID programs'.[2] At US academic institutions a number
of centres have been established with a particular focus on religion, international
development and international relations. They include the Berkley Center's Religion and
Global Development programme at Georgetown University, Washington, DC, which is
interested in the engagement of religious communities around global policy challenges,

and in bringing together stakeholders to examine best practices and advance collaboration.[3] The centre also now houses the *World Faiths Development Dialogue*, established in 2000 by George Carey, the then Archbishop of Canterbury, and James Wolfensohn, then president of the World Bank.[4]

A further example includes the *Luce Project on Religion in Global Civil Society*, a three-year project based at the Orfalea Center for Global and International Studies, University of California, Santa Barbara.[5] Since 2005 the Henry Luce Foundation has been providing substantial grants to fund projects on religion and international affairs.[6]

In Europe also, a shift towards taking religion more seriously in government agendas as well as academic research is noticeable. In the United Kingdom, between 2005 and 2010 DFID funded a large £3.5 million research programme based at the University of Birmingham[7] (Clarke 2007; Stambach 2005). In Switzerland (Holenstein 2005), Scandinavia and the Netherlands such reflection is also under way. In the Netherlands, for instance, the *Knowledge Centre Religion and Development* has been established by the Oikos Foundation, the Inter-Church Organisation for Development Cooperation (ICCO) and the Institute for Social Studies in The Hague.[8]

Understanding and assessing the 'turn to religion' in mainstream development studies, policy and practice

Another key aim of this book, therefore, is to understand why this shift has taken place and its implications for development processes and outcomes, and for religious traditions and their adherents in different contexts. The reasons are complex and it is difficult to pinpoint one or two straightforward causes, but certainly the rise of the religious right in the United States has led to increased funding for FBOs domestically and internationally, both reflecting and fuelling the view that they have a 'comparative advantage' over secular organizations (see Chapter 8). Moreover, following 9/11 in particular, it no longer seemed possible for western governments to ignore or marginalize religion, and correspondingly increased attention has been paid to it in social, political and development studies. In addition, by this time a focus on religion and culture had already begun to emerge in social movements, with the rise of 'identity politics' since the 1980s, which often centred on religion as a source of meaning and social action. Finally, as we shall see in Chapter 2, the shift from the predominantly economics-focused view of development adopted by western governments following the Second World War to a view that included consideration of 'bottom-up development' and 'human development', which sought to measure development in terms of indicators that went beyond the economic, paved the way for this 'turn to religion'.

While many welcome this 'turn', others are more cautious, especially where engagement with religious leaders or organizations may risk undermining progressive development goals such as gender equality (Pearson and Tomalin 2008; Tomalin 2011; Tadros 2010). The 'resurgence of religion' in recent decades has had a marked impact on women's rights globally, and this is something that the international development community needs to be acutely attuned to (Bayes and Tohidi 2001). Moreover, there has been criticism that this newfound interest in religion among development policy makers and agencies is guilty of picking and choosing which types of religion to engage with (Deneulin and Bano 2009). For instance, it has been demonstrated that development donors are more likely to seek engagement with organizations that express their faith 'passively' than those that obviously combine their development work with activities that aim to gain converts (Clarke 2008), and the views of so-called religious extremists (itself a problematic and loaded term) are generally not considered to be consistent with 'achieving development gains' (Haynes 2007: 1).

While all religious traditions have the potential to generate sexist or 'extreme' versions, there is also much activity across different religious traditions that aims to counter such tendencies. Religious traditions are not homogeneous and may include a range of different approaches to belief and practice, from those that are 'extremist', divisive and/or unequal in their treatment of men and women, for instance, as well as those that are the opposite. This diversity within religious traditions, as well as among different religions, presents a dilemma for development actors. Instead of being able to neatly categorize particular religions as having fixed teachings about, or approaches to, areas of concern to development (e.g. education, health or wealth), we find a range of perspectives in different locations, influenced by historical, political, social and even economic factors. Chapter 4 will look more closely at this dilemma and will investigate attempts to characterize and pin down the approaches of different religions to key development concerns, with a particular focus on wealth, poverty and economic development.

Unlike many areas of development studies, this is a new and emerging field of academic study and is not as well mapped as, for instance, 'gender and development' or 'development economics'. While many in the development community are more likely these days to think about religion as a topic that is relevant to their work, reflecting broader social, cultural and political shifts that have thrust religion back into the public sphere, others are less convinced that it is a progressive or even a particularly relevant force. They may also be concerned that there has been a sudden rush to bring religion into development studies and policy without adequate consideration of the consequences. So, this is not only a new field of academic study but also one that deals with controversial issues and does not receive a whole-hearted 'thumbs up' across the international development community. Nonetheless, despite the potential problems and tensions involved, there is a pressing need for a better understanding of the relationships between religions and development, in terms of the ways in which religions potentially impact upon and shape both development processes and development outcomes.

Prior to the 'turn to religion', literature that considered the relationships between religions and development tended to focus upon demonstrating the importance of religion at both the theoretical and the practical levels, arguing for its inclusion in development policies and practices (e.g. Ver Beek 2000; Eade 2002). However, it is now necessary that we engage in more critical reflection on the issues, and produce tools of analysis to deal with the challenges posed by the introduction of religion in this field. The aim of this book is to provide timely, grounded reflection on relevant practical and conceptual issues that need to be taken into account when working with religion in development.

Structure and approach of the book

In Chapter 2 the primary aim will be to understand the reasons why mainstream development studies, policy and practice have ignored or marginalized considerations of religion and culture, as well as why religious and cultural issues are now being given a greater degree of attention. While Chapter 2 will focus on the shifts in development theory and practice that have opened up space for considerations of religion and culture, Chapter 3 examines concepts and theories about religion. These are covered in some detail, but this is necessary since what is typically meant by religion and religious change in the Global North is often poorly suited to approaching similar issues in developing contexts. It is necessary to question the extent to which the familiar definitions, concepts and theories that we bring to the study of religion are adequate for an examination of religions and development.

In the remaining chapters, six thematic sets of questions are addressed (Chapters 4–9). These raise critical discussion points that are currently being debated throughout the international development sector as well as within the broader academic literature. In order to address these questions, the book makes use of three types of material: academic research; material from both 'faith-based' and secular development agencies and organizations; and 'theological' or 'insider' material from within religious traditions. These six sets of questions are not the only ones we could ask but reflect current debates in this area following the recent entry of religion into development discourse:

1 How might members of religious traditions approach 'development'? In what ways does religion influence visions of what counts as development, as well as the policies and strategies employed to reach those visions? Are these religious approaches significantly different from secular approaches? (Chapter 4)
2 Are religious teachings compatible with human rights? What are the implications for 'development'? (Chapter 5)
3 Is religion bad for women's development and the pursuit of gender equality? (Chapter 6)
4 Is it correct to assume that religious traditions teach that one should care for the natural environment? Is this relevant in practice? In what ways might religion contribute towards 'sustainable development'? (Chapter 7)

5 Do 'faith-based' development organizations have a comparative advantage over 'secular' organizations? (Chapter 8)
6 What are the implications of the 'rise of religion' for mainstream development studies, policy and practice in the twenty-first century? (Chapter 9)

The aim of this book is not to cover all topics of relevance, but instead to inform the reader about the relevant practical and conceptual issues that need to be taken into account when working with religion in development. There are therefore gaps in the book in terms of the coverage of topics, since in order to deal with a selection of topics in detail, others have been omitted. Those that will receive minor attention, despite their importance to the field, include health, education, religion and conflict, interfaith activity, concerns about accountability, weak and failing states, governance, indigenous rights, urbanization, the financing of FBOs, and aid harmonization. Nevertheless, it is anticipated that the insights offered throughout the book can be applied to different issues and concerns within the field of 'religions and development', including those that have been omitted.

Neither does the book cover all religious traditions, although it does focus on a good selection. In addition to Christianity, there are case studies and examples that look in some detail at Hinduism, Buddhism and Islam, in particular. The regional focus of the book is similarly broad, with case studies and examples from across the globe. Finally, I adopt an open-minded stance towards both religion and conventional and alternative models of development. My aim is not to promote or endorse a particular approach to 'religions and development', or to favour religion. Instead, the book presents a critical analysis of the aid business, in terms of its failure (at least until recently) to fully acknowledge and engage with cultural and religious diversity, and of the implications of the more recent 'turn to religion' for development and religious traditions.

Summing up

The aim of this chapter has been to analyse the reasons why religion has been ignored or marginalized by mainstream development research, policy and practice and to begin to examine some of the factors that have supported the more recent 'turn to religion'. While many welcome this shift, arguing that it offers the potential for more successful development outcomes, others have questioned the extent to which this engagement with religion indicates genuine dialogue with traditions on their own terms. Instead, there is some suspicion that the engagement with religion reflects the priorities and agendas of mainstream development actors. Others still are concerned that development donors and secular NGOs seem in rather a hurry to engage with religions without proper assessment of the likely outcomes. However, what these debates do suggest is that religion – whether or not it is good or bad for development – is certainly relevant. In the following chapter the aim is to examine the reasons for the 'turn to religion' by mainstream development research, policy and practice in more detail. In particular, we will explore shifts in approaches to donor-led development since the end of the Second World War that have progressively opened up space for consideration of religion and culture.

Questions for discussion

1 In what ways is religion relevant for international development?

2 What have been the principal reasons for the marginalization of religion by mainstream development actors?

3 What is the 'turn to religion' and how has it influenced mainstream development studies, policy and practice?

Recommended reading

To date, there are only a handful of books focusing directly on religions and development. These include:

Clarke, Matthew (ed.) (2013) *Handbook of Research on Development and Religion,* **Cheltenham, UK and Northampton, MA: Edward Elgar Publishing Limited.** This handbook brings together research on religion and development studies, illustrating that as religious identity is integral to a community's culture, exclusion of religious consideration will limit successful development interventions.

Barnett, Michael and Janice Stein (eds) (2012) *Sacred Aid: Faith and Humanitarianism,* **Oxford and New York: Oxford University Press.** This book explores the global humanitarian movement, originating within western religious organizations in the early nineteenth century, as one of the most important forces in world politics in advancing both human rights and human welfare

Deneulin, Séverine with Masooda Bano (2009) *Religion in Development: Rewriting the Secular Script,* **London: Zed Books.** This book argues that development needs to rewrite its 'script', which has been heavily influenced by secular traditions, through engaging with religions on their own terms rather than instrumentally.

Haynes, Jeffrey (2007) *Religion and Development: Conflict or Cooperation?* **Basingstoke, UK: Palgrave Macmillan.** Against the backdrop of rising religious 'fanaticism' across the globe, this book highlights the positive contribution that religions can make to challenges such as conflict resolution, economic development and environmental sustainability.

ter Haar, Gerrie (ed.) (2011) *Religion and Development: Ways of Transforming the World,* **London: C. Hurst.** This edited volume is a more recent addition to a growing literature that explores how religious resources might be harnessed for development.

Tyndale, Wendy (ed.) (2006) *Visions of Development: Faith-Based Initiatives*, Aldershot, UK: Ashgate. Tyndale's book is written from a background of decades of experience working in human rights and development. The chapters in the book focus directly on stories concerning the ways in which different groups and movements in Africa, Asia and Latin America perceive development as linked to religion.

Websites

http://berkleycenter.georgetown.edu/programs/religion-and-global-development The Berkley Center's Religion and Global Development programme is interested in the engagement of religious communities around global policy challenges, bringing together stakeholders to examine best practices and advance collaboration.

http://berkleycenter.georgetown.edu/resources/topics/religion-and-development-database This website provides a database of details about organizations and programmes relevant to the intersection between religion and development.

http://berkleycenter.georgetown.edu/wfdd 'World Faiths Development Dialogue' (WFDD) was established in 2000 by George Carey, the then Archbishop of Canterbury, and James Wolfensohn, then president of the World Bank.

www.religionsanddevelopment.org The Religions and Development (RaD) research programme was based at the University of Birmingham (2005–2010) and was funded by DFID. Of particular interest is the 'publications' section, comprising literature reviews, background papers and working papers detailing research findings.

2 Approaches to the theory and practice of development: from 'estrangement' to 'engagement' with religions

This chapter will:

- outline different meanings or conceptualizations of 'development';
- examine how approaches to donor-led development have evolved since the Second World War;
- explore the implications of shifts in approaches to development for the relationships between religions and mainstream development studies, policy and practice;
- examine different disciplinary approaches to the study of 'religions and development'.

Introduction

The primary aim of this chapter is to help us understand the reasons why mainstream development studies, policy and practice have ignored or marginalized considerations of religion and culture, as well as why religious and cultural issues are now being given a greater degree of attention. There has been a shift from 'estrangement' to 'engagement' (Clarke and Jennings 2008a: 2). We will begin the discussion by asking: when we talk about 'development', what do we actually mean? Are we talking about development as a general process of social or economic change or the more specific western donor-driven development project that emerged after the Second World War? Do we also intend to include forms of charity, philanthropy, disaster relief (action that immediately follows a disaster) and humanitarian assistance (activities taking place in the days and weeks following a disaster) as well as 'development', which is regarded as a more long-term and sustainable process of transformation? While these activities are normally distinguished from development, they are nonetheless conceptually related to it and are, moreover, types of activity that have engaged religions for many centuries. In order to examine the relationship between religions and development, it is important to be clear what sort of 'development' we are talking about.

As was discussed in Chapter 1, the main understanding of development adopted in this book is that of the western donor-driven aid business which emerged after the Second World War. I am interested in undertaking a critical analysis of this business with respect to its failure to engage (at least until recently) with cultural and religious diversity. The previous neglect of religion in mainstream development studies, policy and practice, and the recent increase in interest, have occurred in parallel with a perceived resurgence of religion generally (in both the private and the public spheres, both on the part of individuals and in politics nationally and internationally in many parts of the world). However, it is also the case that approaches to development theory and practice have evolved, thereby opening up space for considerations of religion and culture.

The chapter concludes with a discussion of the ways in which different academic disciplines have approached the topic of religions and development. The newness of this field of study means that in all disciplines there is a distinct lack of literature focusing specifically on 'religions and development'. There is, however, much that is more broadly and directly relevant to this area of study in terms of methodology as well as theoretical and empirical work.

What is 'development'?

In their book on theories and methods in international development studies, Sumner and Tribe delineate three meanings or conceptualizations of 'development':

> The first is historical and long term and arguably relatively value free – 'development' as a process of change. The second is policy related and evaluative or indicator led, is based on value judgements, and has short- to medium-term time horizons – development as the MDGs [Millennium Development Goals], for example. The third is post-modernist, drawing attention to the ethnocentric and ideologically loaded Western conceptions of 'development' and raising the possibilities of alternative conceptions.
>
> (2008: 11)

Development as social change

These three meanings are useful in thinking about the relationships between 'religions' and 'development'. First, religious traditions in different contexts have influenced processes of change throughout history, often without conscious efforts to implement particular agendas. They have impacted incidentally upon cultural practices and social, economic, legal and political systems, as well as technological and scientific progress. At the same time, religions have themselves been influenced by the surrounding culture within which they are situated. Box 2.1 outlines one of the best-known examples of religion facilitating 'development as a process of change': Max Weber's theory of the 'Protestant work ethic' (1930 [1904–1905]). This proposed that Protestantism laid the foundation for the emergence of capitalism in Europe in the eighteenth century.

While much social change is 'non-prescriptive' in nature (that is, societies change over time under different influences without conscious direction by the state or other bodies, including religions), it can also be prescriptive or intentionally brought about through interventions guided by theories about what social changes are considered desirable. It is here that the various theories popular within the western-driven development project since the late 1940s, including modernization theory, dependency theory, basic needs and neo-liberalism, can be located (see p. 25 ff. for a discussion of these theories). Similarly, religions also often consciously guide or seek to influence social change, with religious visions of how things should be actively shaping changes within societies. Much of the discussion in Chapter 4, which examines attitudes within religious traditions to modern development concerns (with a focus on wealth, poverty and economic development), can be located within this understanding of 'development as a process of change'. Most of what religions have to say about these issues pre-dates the emergence of modern 'development', yet today we find these teachings, values and practices being invoked in modern development discourses by religious actors and organizations seeking to alleviate poverty or to call for a more equal distribution of wealth.

Box 2.1

The Protestant work ethic

Weber compared Protestant (Calvinist) Britain with Protestant (Lutheran) Germany and Catholic Italy and Spain to test his view that ideas have an independent and determining role in society. He demonstrated that economic development was greater in Protestant than Catholic countries, arguing that some link existed between Protestantism and capitalism. In his view, the Lutheran view of German Protestantism was not conducive to capitalist development, since the individual was required to accept his or her position in life and to look forward to rewards in heaven. It was the Calvinist doctrine, by contrast, that fostered a this-worldly emphasis upon hard work that had stimulated capitalist growth in Britain. While the doctrine of predestination in Calvinism meant that one could do nothing to change whether or not one was among God's 'elect' (and therefore assured of a place in heaven), hard work leading to worldly success and the accumulation of wealth (which was then reinvested for the purposes of capital growth, since the puritanical values of Calvinism shunned obvious signs of wealth) was considered to be a sign that one had been 'chosen'. Thus, a rationale was given to the accumulation of wealth, which, according to Weber, was one impetus that stimulated the emergence of capitalism. By contrast, when he studied some non-western cultures (e.g. China and India) he found that certain aspects of their traditional cultural systems did not foster an environment where capitalism was likely to flourish. In *The Religions of India: The Sociology of Hinduism and Buddhism* (1958), he was interested in examining the links between the structure of Indian society and its religious systems (e.g. caste and Hinduism), suggesting that the 'spirit' of the caste system, reliant upon the idea of cycles of rebirth and the superiority of the Brahmin class, acted against the indigenous development of capitalism in India.

Source: adapted from Tomalin (2007).

Development as the modern donor-driven development industry (post-1945)

Sumner and Tribe suggest that the second meaning of development is that 'which is likely to be favoured by practitioners within the development community notably in international development agencies' (2008: 13). This understanding of development constitutes the efforts of the 'developed' western nations since the end of the Second World War to undertake specific activities with the aim of bringing about 'development' in less economically wealthy nations. Box 2.2 presents an extract from US president Harry S. Truman's 1949 Inaugural Address, which has been viewed as setting the scene for this new agenda (Sachs 1992).

While the aim of this new agenda was to bring about radical transformations in societies to enable them in the long term to take advantage of the benefits of modernization, thereby eradicating poverty and lessening inequality, early models of top-down development were criticized for carrying out charity rather than engaging with people in ways that could enable them to take their own development forward. Development is often contrasted with 'charity', the giving of aid to alleviate immediate needs, because the latter is considered unlikely to contribute towards long-term and sustainable change. From the 1980s,

---

Box 2.2

President Truman's Inaugural Address, 1949

In a speech that is often considered to mark the beginnings of the modern donor-driven development industry, Truman declared that

we must embark on a bold new program for making the benefits of our scientific advances and industrial progress available for the improvement and growth of underdeveloped areas.

More than half the people of the world are living in conditions approaching misery. Their food is inadequate. They are victims of disease. Their economic life is primitive and stagnant. Their poverty is a handicap and a threat both to them and to more prosperous areas.

For the first time in history, humanity possesses the knowledge and skill to relieve suffering of these people.

The United States is pre-eminent among nations in the development of industrial and scientific techniques. The material resources which we can afford to use for assistance of other peoples are limited. But our imponderable resources in technical knowledge are constantly growing and are inexhaustible. . . .

The old imperialism – exploitation for foreign profit – has no place in our plans. What we envisage is a program of development based on the concept of democratic fair-dealing.

Source: cited in Rist (2002: 71).

bottom-up and participatory approaches to development sought to move away from a charitable model and to embrace participation and cooperation to enable people to help themselves rather than relying on others to provide assistance to them. Charity is also often viewed as carrying a stigma for those who receive it, and therefore to be demeaning and potentially humiliating. As the Uruguayan writer Eduardo Galeano tells us, 'unlike solidarity, which is horizontal and takes place between equals, charity is top-down, humiliating those who receive it and never challenging the implicit power relations' (2000: 312). Nevertheless, charity continues to play an important role, especially in disaster relief and humanitarian assistance, which are important in the aftermath of a natural or human-made disaster but do not necessarily result in development over the long term.

All religions have a long tradition of charitable work, including giving food and other items to the poorest and caring for orphans and the sick and dying. Moreover, the impulse to rush to the assistance of those affected by natural disasters or war is deep-seated (although of course it is not only religion that provides an impulse to assist victims of war and disasters). While much of the activity that religious organizations have traditionally undertaken would seem to be charity or disaster relief rather than social or human development, religious organizations have also been important providers of education and health, for instance, as well as involved in activities to provide clean drinking water, strengthen livelihoods and secure human rights. Most religious traditions have mechanisms for helping the poor as a central feature – for example, *dana* ('selfless giving') within Hinduism and Buddhism, and *zakat* within Islam (the injunction to tithe a portion of one's wealth every year for charitable purposes). However, the distinction between 'charity' and 'development' is not always clear-cut. People in many religious traditions do not only give charity. Instead, they also ask questions about why poverty and inequality exist and what must be done to overcome them. In asking these questions, the gap between charity and development is bridged. From a religious perspective, it is humans' greed and selfishness and their failure to act in accordance with the teachings of the religious tradition that cause poverty and suffering. Charitable practices are therefore to be seen as reflecting the way in which humans would act in an ideal future society; they are practices that are forms of 'spiritual training' for the devotee to cultivate the correct ways of being and acting. For instance, for Hindus and Buddhists *dana* must be done without attachment to the results of the action, that is to say without any personal consideration of the benefits. The aim is to cultivate the mindset or attitude that would be the norm in a 'developed' society. Charitable practices are thus embedded within broader models for how societies should be – 'developmental models' – although this is not always evident in practice, and religious models of charity often do not appear to reflect such deeper visions of transformation.

As will be demonstrated in Chapter 4, these visions of religious charity and social change are very ancient and pre-date the modern post-Second World War development project. However, over the past decades increasing numbers of religiously based organizations have emerged or shifted their own agendas in order to meet the requirements of the 'development industry', and members of religious traditions increasingly engage with the language and concepts of modern development.

Development as a dominant discourse of western modernity: the post-development view

The third conceptualization of 'development' suggested by Sumner and Tribe is also relevant to an examination of the relationships between religions and development. The idea that 'development' is a dominant discourse of western modernity has been seized upon by 'post-development' critics, who draw attention to the imposition of western ethnocentric visions of development upon the Global South by the Global North, as well as arguing that in many instances 'aid' has actually increased inequality and poverty. This view considers that there is much more continuity between colonialism and the development industry than development studies has so far chosen to acknowledge. As Kothari, a post-development thinker, writes, 'there has been a political imperative to distance the international aid industry from the colonial encounter so as to avoid tarnishing what is presented as a[n] humanitarian project far removed from the supposed exploitation of the colonial era' (2005: 51). A post-development perspective does not necessarily reject positive social change, but rather rejects the linking of this to the mainstream idea of 'development', which is associated with the global aid business.

One aspect of this 'post-development' critique has been the view that the development agenda imposed by the Global North is colonialist because it ignores and/or overrides local cultures and religions. This critique suggests instead that, rather than an approach to development which is modelled on western values and culture, and which assumes that western-style economic prosperity and modernization is the only way to 'develop', social change and progress should reflect local histories, cultures and religious traditions. While the global aid business (which promotes the second understanding of 'development' outlined above) does now appear to be engaging with religion and culture to a greater degree than in the past, scholars and practitioners who are sympathetic to the view that development is a top-down western hegemonic project are critical that this 'turn to religion' does not go far enough. Instead, they suggest, mainstream development actors seem to pick and choose types or parts of religion with which to engage that resonate with their underlying worldview (Deneulin and Bano 2009). By contrast, a 'post-development' approach to 'religions and development' would require that religions are treated 'holistically' rather than being forced to bend and adapt to northern development priorities. From a post-development perspective, a genuine engagement with religions would probably challenge many of the underpinnings and priorities of current dominant global development paradigms, giving rise instead to approaches based on the promotion of localized, culturally appropriate socio-economic systems that incorporate religious values and recognize the importance of the 'spiritual' in people's lives.

While such an approach could present a radical challenge to existing development policies and outcomes, and reflect more closely the values and priorities of different communities, we cannot assume that religious worldviews are necessarily the most effective drivers of positive social change, because there is sometimes a tendency towards romanticism in post-development thinking that valorizes the social, economic and religio-cultural traditions of pre-industrial communities. Nor can we assume that those holding a religious worldview will necessarily reject modern economic theory and practice, such as neo-liberalism (see pp. 39–40 for a discussion of Rudnyckyj 2011). While important in highlighting the nature and role of development as a western discourse, therefore, the post-development perspective does have a number of problems attached to it (Ziai 2004). These include a tendency to overstate the destructiveness and ethnocentricity of development models from the Global North; a bias towards cultural relativism, which can be dangerous for women and minorities where traditional cultures do not place a strong emphasis on equality; a romanticization of the role that the cultural and religious traditions of pre-industrial communities can play in promoting empowerment and integrity compared to western secular frameworks such as human rights; and an assumption that local traditional models of socio-economic change will be favoured in the Global South, rather than modern economic theories and systems.

* * *

All three understandings or conceptualizations of development are relevant to this book (but the focus will be upon the second). At the heart of all religious traditions is a conception of human and social progress, and each offers ways of achieving this, as well as blocking change when it is felt to go against core teachings. However, in the past, and often today, this conception of progress is not called 'development' and it may not directly or obviously connect with modern development discourse. As has already been noted, places of worship and 'faith-based organizations' have traditionally engaged in various forms of charity, relief work and helping the poor and sick, and today such activities continue (Plates 2.1 and 2.2). In addition, religious values have played a role in shaping social attitudes and forms of politics; religion has an ongoing relationship with forces that seek to maintain ethnic, class and gender divisions; and religions may promote or resist change. These may not map onto modern concepts of development popular in the Global North. However, nowadays members of religious traditions and 'faith-based organizations' often engage with development donors and talk about how their teachings and practices support or offer alternatives to secular understandings of development. It is therefore important for those engaged in this dialogue to understand not only how religions in the past and today conceive of human progress and their responsibilities towards the poor, but also how western understandings of 'development', and how it can best be achieved, have evolved.

Plate 2.1 *Tibetan Buddhist monks aid rescue effort in Kyegundo, 20 April 2010*

Source: STR/AFP/Getty Images.

Tibetan Buddhist monks are pictured handing out bottled water from a relief truck following a powerful 6.9-magnitude earthquake that struck on 14 April 2010 in Kyegundo, eastern Tibet (Yushu Tibetan Autonomous Region, Qinghai Province), killing thousands and leaving 100,000 homeless.

Plate 2.2 *A patient with AIDS prays for her boyfriend, August 2006*

Source: AFP/Getty Images.

This AIDS patient is 'praying' for her boyfriend, who died at a temple for patients in Thailand's northern Lopburi province. The temple is built at the foot of a small mountain, 150 km north of Bangkok and, at the time the photograph was taken, was home to 550 AIDS patients, including 140 children.

How have approaches to development evolved?

There have been important changes in approaches to development since the Second World War that, according to Deneulin and Rakodi, have implications for 'the conceptualization of the relationships between religions and development' (2011: 46). However, there is disagreement in the development studies literature not only about the precise definition of 'development' but also about when it is thought to have emerged. As we have seen, Kothari (2005) and other post-development thinkers are critical of those who locate the origins of development in the 1940s, following the Second World War, and instead consider that it is a continuation of colonialism. While development studies does engage with this colonial history in terms of explaining the historical circumstances that pre-date and shape the global political landscape from which development emerges, it is less likely to consider that the activities and objectives of development are coterminous with the colonial spirit.

The colonial period

From the point of view of understanding the relationship between religions and development research, policy and practice in the contemporary era, an exploration of colonial attitudes towards both religion and processes of economic and social change is important. Chapter 3 will examine the ways in which western, and by implication colonial, understandings of what a religion is shaped views of the religious traditions encountered in the colonies. Not only did the presence of Christian missionaries have an impact upon indigenous religiosities, but the very model of what a religion was considered to be heavily influenced the ways in which the West has studied and understood non-western religions since (Smith 1991 [1962]; Beyer 2006; Asad 1993).

In terms of social and economic change, the colonial enterprise was very much guided by the emergent theories of classical political economy in the late eighteenth century, which, as Nkurunziza points out, 'presents the first formulations of the conception of development, more commonly referred to at the time as "progress"' (2007: 2). Theorists such as Adam Smith, David Ricardo and Karl Marx, in different ways, considered that imperialism and colonialism were essential for capitalism to flourish. According to Nkurunziza, by the 1940s 'the theory of imperialism also began to adjust to the fact that the postulated industrialisation in the colonised territories was not materialising' (ibid.: 3), and political independence was increasingly viewed as the means to bring about economic growth. In fact, on the whole, only the 'settler colonies' had been industrialized, with India, for instance, subjected to a dismantling of its well-developed textile industry as one was established in Britain, which relied on India only for its raw materials.

Following the Second World War, dozens of countries in Africa, Asia and the Caribbean joined already independent, although still 'underdeveloped', Latin

American countries (Haynes 2008: 7–8). Concerns over the future of these countries, coupled with the fact that the world depression of the 1920s and 1930s had already dampened the optimism of 'blind faith' in the market, led western governments to agree that there was a need to provide 'development aid' to kick-start the economies of 'underdeveloped' nations. To replicate the success of the Marshall Plan in rebuilding the economies of Europe after the Second World War, foreign aid to and investment in poor countries was promoted in the belief that as their economies grew, the economic benefits would 'trickle down' to the poorest communities. This provided the justification for the beginnings of bi- and multilateral financial institutions, especially the so-called Bretton Woods institutions set up in 1945 (the International Monetary Fund (IMF) and the World Bank – see Box 2.3), which have come to be known as the 'aid business' (Rakodi 2011: 30).

Box 2.3

The International Monetary Fund (IMF) and the World Bank

The 'aid-business' had its genesis in 1945, when

[a]t a mountain resort in New Hampshire [in July 1944] called Bretton Woods, financial representatives from the 44 allied nations devised institutions to alleviate the impediments to international financial growth that had arisen as a result of the war. The International Monetary Fund (IMF) was created to restore the volume of international trade that had dropped due to instability since the 1930s. . . . A pool of currencies would be contributed by member states from which any member country could draw upon in order to correct any balance of payment problems. . . .

In December 1945, the IMF and The International Bank for Reconstruction and Development, also known as the World Bank, were officially established. The original purpose of the World Bank was to grant loans to rebuild Europe after the war. Both institutions worked in conjunction with the Marshall Plan for the redevelopment and economic stabilization of Europe.

Source: www.munfw.org/archive/45th/ecosoc3.htm (last accessed 11 September 2012).

Modernization and economic growth (1940s–1960s)

As President Truman's 1949 address indicates (Box 2.2), the emergent aid business was underpinned by a firm belief in modernization, an agenda to spread the benefits of modernity enjoyed by the West to poor countries. This view came to dominate development discourse in the 1950s and 1960s. While economic growth was being given a kick-start, modernization was considered to be an inevitable and irreversible trajectory. As Melkote and Steeves write:

Modernization is based on liberal political theory and is therefore grounded in the grand project of 'Enlightenment', namely reasoning, rationality, objectivity, and other philosophical principles of Western science. . . . Implicit in the discourse of modernization is a certain philosophy of what development in the Third World should be, and how it should be brought about.

(2001: 71)

Modernization theory was developed in the 1950s and early 1960s by social scientists, mainly in the United States, prominent among whom was Talcott Parsons (1902–1979). For Parsons and others, the United States was considered to be the leading modern society, marked by 'democratization, industrialization, and the expansion of education and literacy' (Dillon 2010: 168). However, the extent to which modernization and westernization were the same thing were debated (and still are), with Shils, for instance, a contemporary of Parsons, arguing that 'we should . . . avoid the term "Westernization" because it suggests something attached to the West' (Gilman 2007: 143). Nevertheless, according to Gilman, 'the one thing Shils was sure of . . . was that modernization meant to become "like the West" but (in principle) without the subordination that this implied' (ibid.: 143).

Central to the theory of modernization is a distinction between traditional and modern societies, and it was assumed that all traditional societies would eventually undergo industrialization, urbanization, secularization and rationalization, as had occurred in much of Europe and North America. These are key features that sociologists identify as characterizing the period known as 'modernity', which began to emerge as early as the sixteenth century with the European Renaissance and the Protestant Reformation, coming to fruition with the rise of science, the Enlightenment and the Industrial Revolution (Giddens 1991; Berman 1982). While Truman's speech indicates a very positive view of modernity and its benefits, since the emergence of the discipline of sociology in the late nineteenth century, scholars have also drawn attention to the problems and costs associated with modernity. Marx, for instance, saw it as an era of devastating exploitation of the working classes, with the shift to modernity being marked by the rise of capitalism; for Durkheim, the loss of a overarching moral vision binding society together, caused by the division of labour and rapid social change, resulted in 'anomie' ('normlessness'); and for Weber, the increased rationalization of society was dehumanizing, 'imprisoning us in a featureless cage of bureaucratic routine' (Giddens 1991: 137).

As Nkurunziza writes, modernization theory promoted the view not only that 'an inevitable outcome of economic change would be social transformation' but also that 'a reorganisation of social and cultural institutions would *facilitate* economic growth' (2007: 6; emphasis in the original). The question for the emergent development industry was how to accelerate this process through various interventions. The problems of developing countries were considered to be essentially economic, with economics dominating thinking about development and development policies and economists designing and implementing macro-level five-year plans in most newly independent countries. Because, according to Melkote and Steeves, 'participatory or autonomous development by local communities was considered slow, inefficient, and more often than not, unlikely' (2001: 72), development was to be achieved by an injection of external funds to overcome the main perceived constraint on economic growth: developing countries' shortage of capital. In addition to the injection of capital and the commercialization of agriculture, bilateral and multilateral aid organizations provided governments with funds to train entrepreneurial elites, expand education, make strategic use of the mass media to promote development ideas among their populations, and promote democracy and modern legal systems (Nkurunziza 2007: 8).

From the very 'beginnings' of this modern development project, underpinned as it was by modernization theory, religion and culture were given no place of importance. Local cultures and religions were not seen to be relevant to this grand modernizing plan; in fact, they were considered to be obstacles to economic development. In addition, it was assumed that religion would become less important to people as they experienced economic development. The secularization of Europe as it had modernized was taken as a blueprint for what would happen elsewhere.

The main criticisms were of the idea that modernization equals westernization, and that this approach to economic development is a form of western colonialism and imperialism. As Dillon points out, many neo-Marxist critics at the time (e.g. Frank 1967; Cardoso and Faletto 1979) also 'contended that modernization theory was essentially ethnocentric because of its presumption of American society as the prototype and any societal deviations from it as inferior' (2010: 169). While the neo-Marxist critics of modernization theory were less likely to draw attention to its failure to acknowledge the importance of religion and culture, promoting instead Marxist frameworks of analysis, they were nonetheless influential in beginning to lessen the monopoly of modernization theory over development thinking.

Dependency and underdevelopment (1960s–1970s)

By the late 1960s there was little sign of convergence between the developing and the developed world; in fact, the gap was increasing not only between developed and developing countries, but also between rich and poor in developing countries. There had been economic growth in some developing countries, but the so-called trickle-down effect did not seem to be working. Moreover, modernization theory was criticized for its simplistic distinction between traditional and modern societies, as well as the view that modernization essentially meant that all countries would and should become like the West.

Modernization theory came under attack in the late 1960s and early 1970s, mainly from a new approach called 'dependency theory', which, influenced by Marxism, sought to explain why developing countries remained 'underdeveloped'. It was largely a product of Latin American scholarship, later popularized in the West through the work of individuals such as Andre Gunder Frank, and sought to explain why, despite political independence, Latin American ex-colonies had been unable to achieve development (e.g. Amin 1976, 1992 examines African examples; Conway and Heynen 2008: 92). While it did not reject the idea of modernization per se, it did not believe that capitalism was capable of achieving it. Sharing much in common with earlier theories of imperialism, which argued that the colonies in the 'periphery' provided cheap labour and resources for the growth of profit in the 'core' colonizing countries, developing economies in postcolonial contexts continue to undergo such exploitation, reflecting 'the adverse terms of their incorporation into the global economy, terms they cannot change' (Nkurunziza 2007: 12). Thus, according to Sylvester, poverty and underdevelopment prevail not only because historical processes had 'stripped colonies of resources, reorganised their lands, pauperised their labour, and created parasitic elites – all so western countries could have and sustain the once-ever industrial revolution' (1999: 706), but also due to the ongoing pseudo-colonialism of the modernization approach to development. Economic development in these 'dependent' or 'peripheral' economies is prevented by the very system they are embedded within. Suggested ways of overcoming this dependency included finding ways for developing countries to increase their own industrial capacity as well as promoting 'revolutionary class struggle and even guerrilla warfare' (Nkurunziza 2007: 13).

Despite its popularity, dependency theory did not manage to oust modernization approaches to development. It was perceived by many to be too radical and abstract (consisting of a rejection of the capitalist model), and was also criticized for failing to account for capitalist development in some parts of the developing world and for viewing elites and governments in developing contexts as overly passive. Nonetheless, it is an account that does still have a resonance today, with Conway and Heynen, for example, writing that 'just as the imperialism of old imposed colonial regimes that fostered dependency and underdevelopment, modern globalization of the post-1980s has several salient features that are de facto successors to these imperial mechanisms' (2008: 95). The inequality of the global capitalist system penetrates development debates today: it is at the core of the Jubilee 2000 'drop the debt' campaign, the Millennium Development Goals and the more recent Occupy movement.

As we will see in later chapters, concerns over global inequalities have attracted much comment and social action from religious actors, who have been strongly involved in initiatives such those mentioned above. However, the extent to which these religious responses employ a neo-Marxist or dependency theory narrative is likely to be compromised by the fact that, as Box 2.4 illustrates, within dependency theory itself there was no place for religion. Marx's famous characterization of religion as 'the opium of the people' meant that religion was seen as a means to control and console the poor rather than as having any potential to fundamentally liberate them. Nonetheless, Marxism and religion have at times come together, with liberation theology being an example of a religious movement that many consider combines Marxism and Christianity (see Chapter 4).

Box 2.4

Marx: religion as the 'opium of the people'

For Marx, religion played a crucial role in his theory but as a response to the social alienation inherent to the capitalist system rather than as a positive social force; this is expressed in his now famous depiction of religion as the 'opium of the people'. Marx considered that social change was the product of material forces rather than any extra-human agency. Throughout the course of human history the differing relationships that people have established with the material world are reflected in the types of social relations they engage in. In this current capitalist phase the working class is exploited by the upper class for the purposes of producing excess capital. Thus, for Marx, when people seek solace in religion they can compensate for their sense of alienation and suffering: it masks and mitigates the true nature of the exploitative structures of capitalism that oppress the working class. Thus, it acts as an 'opiate', blinding people to the real source of their oppression and serving the interests of the ruling or upper classes by enabling them to retain their power and status. Secularization is therefore desirable and, because religion would wither in a socialist society, it is also inevitable.

Source: adapted from Tomalin (2007).

Alongside the emergence of dependency theory, there were other attempts to explain the failure of poor countries to develop, despite economic aid, as well as efforts to redefine development and to devise better development indicators. The 'redistribution with growth' approach argued that while some developing countries had witnessed overall economic growth, this had not been distributed evenly and there needed therefore to be more effort to redistribute wealth, through investment in neglected sectors such as agriculture, education and health (Chenery *et al.* 1974). However, this approach was criticized for failing to challenge the basic inequalities inherent in the capitalist system, instead working within an economic framework that was not capable of fundamentally transforming economic relations. Others argued for a more direct 'basic needs approach' (BNA) that would emphasize meeting the basic needs of all as an end in itself, rather than merely aiming to increase productivity and economic growth. What we find at this time is the beginnings of a critique of the mainstream development agenda for focusing too strongly on economic growth at the expense of other things. A key figure in these 1970s debates was Mahbub ul Haq, a leading World Bank economist who played a crucial role in advocating for a shift away from a primary focus on economic growth and instead towards poverty and social development (1976: 92).

Neo-liberalism and the 'Washington consensus' (1980s)

While these critical voices in development were influential in shaping approaches that have shifted the focus from economic growth – for instance, the 'capabilities approach' and 'human development' – their impact was temporarily truncated in the 1980s during the debt crisis in the Global South (1981–1982), which arose from increasing interest rates on the loans developing countries had taken on and was precipitated by the increases in oil prices in 1973 and 1979 (Simon 2008: 86). In 1981, Brazil, Mexico and Poland announced that they could not service their official debts. Instead of considering writing off all or some of the debt, northern creditor governments and the international agencies, especially the IMF and the World Bank, in a bid to protect international financial systems and maximize 'the prospects for, and amounts of repayment by, debtor countries' (ibid.: 87), forced them and many other developing countries to adopt a set of policies that came to be known as 'structural adjustment'.

Development theories since the Second World War had placed the state at the centre of the development process, reflecting the existence of strong welfare states in many western contexts, as well as the control of the state over economic development in many developing countries. Already in the Global North, by the late 1970s, there had been a resurgence of neo-liberal economics, which revealed scepticism about the role of the state in the economy and instead shifted to what Simon calls a 'rather naïve belief in the "magic of the market"' (2008: 86). By the 1980s the debt crisis provided an opportunity for neo-liberalism to be transferred to the Global South, as an approach later termed the 'Washington consensus' (led by the US government, the IMF and the World Bank) came to dominate international fiscal policy.

The main tool of the Washington consensus was so-called structural adjustment programmes (SAPs), 'neo-liberal' policies implemented by the World Bank and the IMF that aimed to reform governments in developing countries, in particular reducing the extent and scope of state intervention, especially in the economy, as well as promoting engagement in international trade through economic liberalization (Simon 2008: 87). The main accrual of debt pre-dated the SAPs, although many critics also argued that the policy package failed to increase the rate of economic growth, actually increased debt and dependency, and impoverished substantial numbers of people in developing countries whose governments were implementing the programmes.

Since the early 1980s, in the United Kingdom, for instance, a variety of organizations and individuals formed the Debt Crisis Network (similar networks existed in other countries), which campaigned for debt cancellation. In response, a new initiative was launched by the IMF and the World Bank in 1996: the Heavily Indebted Poor Countries (HIPC) scheme. As the website of the Jubilee Debt Campaign explains, while this aimed to 'reduce the external debt of eligible countries as part of a strategy to achieve debt sustainability'[1]

> [t]he original HIPC initiative was widely criticised for providing too little relief, too late: calls which were echoed later in the debt campaign. At this point it was clear that a socially broader-based, more international debt campaign was needed to press home these concerns. . . . The HIPC initiative, while a big theoretical step forward, was slowed down by the IMF's insistence that debtor

governments would have to 'perform' against a set of conditions laid down by the IMF.[2]

By the mid-1990s, continuing concern over the debt accrued by developing countries and the slow implementation of the HIPC initiative led to the emergence of a broader-based and international civil society-led movement, and 'the Debt Crisis Network transformed itself into the Jubilee 2000 Coalition'.[3] As Box 2.5 demonstrates, religious actors and organizations were prominent within these campaigns (see also Donnelly 2002; Plate 2.3). The call for the debts owed by developing countries, including the substantial loans accrued prior to the SAPs, to be cancelled by the start of the new millennium put further pressure on the IMF and the World Bank. While the HIPC-1 initiative had been introduced in 1996 to provide forty-one 'debt-distressed' countries (thirty-three of which were in Africa) with '"an exit strategy" towards "sustainable" levels of debt' (Katsouris 1999: 11), by 1999 only four countries had benefited and only three more were due to benefit by 2001. The slow progress had been due to the difficulty of qualifying as a HIPC and the existence of a six-year qualification period following structural adjustment, to establish a track record in complying with the external conditions. HIPC-2 was agreed at the G7 summit in Cologne in 1999, with the aim of making it easier for countries to qualify for debt relief. However, for the debt campaigners this also failed to adequately address the issues. According to Katsouris, 'the G7 proposals fall well short of debt lobbyists' recommendations', particularly since 'Jubilee 2000 and others called for a one-off cancellation of all HIPC debt' (1999: 14).

Box 2.5

The Jubilee 2000 campaign

Jubilee 2000 was an international coalition in over forty countries, involving religious groups and leaders, particularly Christian ones, alongside trade unions, aid agencies and other campaigning groups. In fact, the very concept of a 'Jubilee' is itself biblical, based on the following passage from Leviticus, a book in the Old Testament: 'Consecrate the fiftieth year and proclaim liberty throughout the land to all its inhabitants. It shall be a jubilee for you; each one of you is to return to his family property and each to his own *clan*' (Leviticus 25:10).[4]

By October 1997 the Jubilee 2000 Coalition had formed and a mass protest was held to coincide with the Birmingham G8 Summit on 16 May 1998. According to the website of the campaign:

> Christian Aid, CAFOD, WDM and TearFund had campaigned enthusiastically behind the Jubilee 2000 campaign, and ensured that almost every available church hall was used to organise a Jubilee 2000 meeting. Their staff and volunteers organised and spoke at meetings, distributed leaflets; wrote articles; mobilised petitions, staffed stalls, and chained themselves to railings! All the while, educating, educating,

educating. By the time the big day – 16 May 1998 – arrived, a huge swathe of the British public, and a fair section of the media, had been taught the basics of international debt and finance. To the astonishment of politicians, and to the surprise of the G8, millions had a firm grip on the issues.[5]

In 2001 the Jubilee 2000 Coalition became known as the Jubilee Debt Coalition and in 2009 launched its Multifaith Project, which aimed to 'strengthen the faith call for debt justice by bringing our small voices together to make a loud appeal out of shared faith values. This is a matter of justice, we are still being heard'.[6] While the website of the project has sections on the approach of different faith traditions to the issue of debt, what is particularly interesting about this initiative is the idea that the faiths share a sense that the debts burdening poor countries and preventing them from reducing poverty are unjust.

Plate 2.3 *Jubilee 2000 group protest to seek the cancellation of all Third World debt in 1998 at the G8 Summit in Birmingham, UK*

Source: © i4images rm/Alamy.

Protestors are pictured holding Christian Aid banners.

The rolling back of the state and the rise of the NGO

The success and perseverance of coalitions such as Jubilee 2000, and others such as Make Poverty History, are in large part due to the existence of a strong global civil society underpinned by networks of non-governmental organizations (NGOs). Within this diverse category of organizations we find a variety of types, ranging from those that are large, international and typically based in the Global North, to smaller, local NGOs in developing countries. Some NGOs focus on advocacy, whereas others are involved in community development or are predominantly service providers. In the Global North, in particular, many NGOs have also developed the capacity to undertake development research, reflecting the view that development is not just about the flow of aid; structural issues need to be understood and addressed, requiring empirical study and analysis.

With the rolling back of the state, which included the privatization of previously state-provided services, the role of non-governmental organizations increased significantly, with 'NGOs fill[ing] gaps left by the privatization of state services as part of a structural adjustment or donor-promoted reform package' (Desai and Potter 2008: 500). The rolling back of the state that is associated with neo-liberal economics, and was built into the SAPs, encouraged and promoted the growth of NGOs, which were promoted as alternatives to the state as service providers and development organizations, as well as vehicles through which development aid could be directed instead of national governments. Just as national governments were perceived to have failed to stimulate economic growth, therefore seeming to justify the increased emphasis on the market, neither could they be trusted to use aid wisely. As a result, the NGO sector experienced massive growth, supported by donors in the Global North. NGOs were perceived to be less corruptible and more likely to be genuinely committed to helping achieve development aims than governments (Lewis and Kanji 2009). As Desai points out, NGOs were also seen as 'more administratively flexible, closer to the poor, innovative in problem solving and more cost-effective than corresponding state partners' (2008: 525). Donors also viewed them as playing an important mediating role between donors and states, and considered that they had the potential to contribute towards the strengthening of civil society and democracy or good governance (Mercer 2002).

Many continue to believe that NGOs have these advantages, but there have also been concerns that many of their advantages are assumed rather than proven or that their 'inherent' advantage may be worn away by increased funding and professionalization (Desai 2008: 528). As Desai writes:

> Increasingly, questions have been asked. Can NGOs deliver all that is expected of them? Is the glowing image realistic? How effective are NGOs? There seems to be more concentration on success stories such as the Grameen Bank and Seva, and a gap appears to be emerging between rhetoric and practice, which raises issues of objective monitoring and evaluation of NGOs' projects, effectiveness, legitimacy, performance and accountability.
>
> (ibid.: 529)

Concerns have also been expressed about an 'NGO-ization' of development, with a fear that NGOs have 'captured' the development effort. If this is the case, it could be problematic in number of ways, including the transmission of development visions by international NGOs based in the Global North, many of which are funded by northern governments and other organizations; a reliance on organizations that may be lacking in both capacity and accountability; and evidence to suggest that in some developing contexts an industry has grown up around NGOs which has more to do with creating jobs and incomes for a comfortable elite than with carrying out development.

While many NGOs are 'secular', others are linked to religious traditions in different ways. These are most commonly referred to as 'faith-based organizations' (FBOs) (although different terms have been used to describe them, including 'religious NGOs' (RNGOs) or 'faith-based development organizations' (FBDOs)). The history and characteristics of FBOs will be discussed in detail in Chapter 8, where it will be suggested that a range of assumptions are often made about these organizations that have not always been adequately tested by empirical research, but which are used to support the view that FBOs have a comparative advantage over secular NGOs. These include the idea that they are closer to the grassroots, more efficient and less likely to squander financial resources, more likely to be trusted, and also that they pay attention to the spiritual or affective dimensions of people's lives, which are important to broader or holistic understandings of development. It is striking that most of these qualities are also considered to characterize secular NGOs, but with these the comparative advantage being claimed is with respect to the state or for-profit organizations (that latterly have taken over many development services following structural adjustment).

The 'post-Washington consensus' and human development (1990s–2000s)

By the end of the 1990s there had been a shift towards thinking about development in terms of poverty reduction and inequality, picking up the undercurrent from critical approaches in the 1970s and reacting to the dismay that SAPs had not achieved their desired outcomes. While acknowledging that some of the reforms included in SAPs were necessary and some have had positive outcomes, the so-called post-Washington consensus emerged, which emphasized the shortcomings of blind faith in the market, aimed to bring the state back in – the 'regulating state' – and recognized that many early structural adjustment (SA) policies had increased rather than reduced poverty. It was acknowledged that countries could not wait for the restoration of economic growth and trickle-down to reduce poverty, but had to both redesign some SA policies and introduce active poverty reduction initiatives. Thus, by the late 1990s 'locally prepared' Poverty Reduction Strategy Papers (PRSPs) were introduced as part of the conditionality for debt write-offs and new loans or grants to HIPC countries. Many SA policies were incorporated into PRSPs, and also many countries without PRSPs continued to implement SA policies. As Simon suggests, this shift to Poverty Reduction Strategies fitted well with donors' reinvention of development as poverty reduction or elimination, and the rebranding of development assistance – previously referred to as development aid – as 'partnerships' in the form of development cooperation (2008: 90).

The adoption of a conception of development that did not focus primarily on economic growth and the renewed emphasis on poverty reduction were supported by evolving development thinking. Box 2.6 describes the emergence of the idea of 'human development', an approach that has become central since the early 1990s, influenced by the work of Paul Streeten *et al.* (1981), Mahbub ul Haq (1995) and Amartya Sen (1999) (Hulme 2007: 2).

'Human development' is understood as a process of enlarging people's choices and enabling them to live happy and fulfilling lives: it considers human beings to be the ends as well as the means of development (Hulme 2007: 2). As an approach to development, it has been highly influential, most notably through the United Nations Development Programme's Human Development Index (HDI), which since 1990 has provided an alternative to per capita GDP for ranking countries according to their level of development. Rather than being based only on monetary measures of wealth and poverty, it is made up of indicators that reflect well-being: a long and healthy life (life expectancy at birth), knowledge (expected years of schooling of children, mean years of schooling of adults) and a decent standard of living (gross national income per capita). While it has tended to be seen by some as an 'add-on' approach to more economically focused models, more recently it has become a key factor driving the work of development professionals (Haynes 2008: 13).[7]

A strong influence on the emergence of human development was the 'capabilities approach' of Amartya Sen, which measures development in terms of 'human capabilities' – the things that people can and cannot do in life. The aim or goal of development should be to enable people to realize these potential

---

Box 2.6

Human development

According to Haynes (2008: 12) the concept of 'human development' can be understood in different ways:

> The notion of "human development" can be understood in various ways. Politically and economically, human development is concerned with stability, security and citizens' relative prosperity. Socially, it relates to literacy, education, social relationships and, more vaguely, "the quality of life". Morally, it involves development of conscience, moral awareness, and the will and associated capacity to act according to societal and cultural knowledge of what is judged to be right – and in the developing world this often overlaps with life's religious and spiritual dimensions. Finally, psychologically, human development is to do with mental health, self-esteem, success in significant relationships, and human happiness.

capabilities. Sen argued that rather than measuring income or consumption, poverty should be measured using indicators of the freedom to live a valued life (1985, 1999). While Sen has been reluctant to formulate a list of capabilities, the North American moral philosopher Martha Nussbaum has suggested a list of ten basic capabilities relating to the following: life; bodily health; bodily integrity; senses, imagination and thought; emotions; practical reason; affiliation; other species; play; and control over one's environment (2000: 77–80). This is a remarkably holistic list that has resonances with the spirit and detail of many examples of religious values. The move to place the human being at the centre resonates with the critique that many religious traditions have had of the focus on economics within donor-led development since the Second World War. As will be demonstrated in Chapter 4 (which will examine views about wealth, poverty and economic development in different religions), although people need material things, including money, to live well, these should not be pursued at the expense of other human dimensions or at the expense of other people. Moreover, Nussbaum clearly states that religious freedom and expression are aspects of what it is to live a good human life. She recognizes the importance of religion as a source of ultimate meaning, and while she acknowledges that it can act against people's human rights, she also suggests that contesting voices within traditions have the capacity to transform religions into important avenues for social action (2000). She is critical of those who dismiss religion for being essentially and irredeemably patriarchal, authoritarian and backward-looking.

The Millennium Development Goals

As is outlined in Box 2.7, the emphasis on poverty reduction and human development is clearly reflected in the current focus of development cooperation on the UN Millennium Development Goals (MDGs): eight international development goals that all 192 United Nations member states and at least twenty-three international organizations have agreed to achieve by 2015.

What we also notice here is that the focus on economic growth has shifted to include other aspects of human life and well-being as significant. The human being is far more at the centre here than in earlier approaches of development. Moreover, the World Bank in its *World Development Report 2000/2001* considered that for the MDGs to become a reality, there needed to be a greater involvement of local people and organizations in decision-making processes, and '[w]hile the Bank did not specifically mention specific organizations in the . . . *Report* there was a clear inference to its recommendations for better developmental outcomes: to achieve the MDGs in the short time-span allotted – just 15 years – would require utilization of all currently under-used human resources, including those of both secular and faith-based organizations' (Haynes 2008: 38; see also Clarke 2007: 80; Alkire and Barham 2005; Boehle 2010). One of the earliest initiatives was that of Religions for Peace, founded in 1961 and currently the largest coalition of religious representatives. It produced a document called 'Faith in Action: Working toward the Millennium Development

---

Box 2.7

The Millennium Development Goals

The MDGs are comprised as follows:

    Goal 1: Eradicate extreme poverty and hunger

    Goal 2: Achieve universal primary education

    Goal 3: Promote gender equality and empower women

    Goal 4: Reduce child mortality

    Goal 5: Improve maternal health

    Goal 6: Combat HIV/AIDS, malaria and other diseases

    Goal 7: Ensure environmental sustainability

    Goal 8: Develop a Global Partnership for Development

Source: www.undp.org/mdg/basics.shtml (last accessed 13 December 2011).

Goals: An Action Toolkit for Religious Leaders and Communities' (Religions for Peace 2005). Not only have religious traditions responded to the MDGs, but development donors have also invited the input of faith leaders.

In contrast to this optimistic account of religious engagement with the MDGs, however, others remain critical that, overall, religions may do more harm than good, for instance in their often poor record on women's rights or Catholic objections to the use of condoms in the fight against HIV/AIDS. However, whether or not religions positively contribute towards the MDGs or human development cannot be answered 'yes' or 'no'. Arguably, we should not be asking questions about whether or not religion is positive or negative for development, but instead about whether or not it is relevant. The answer to the former is likely to become mired in attempts by factional interests, whether these are religious, political or intellectual, to advance their own viewpoint or cause. As this book seeks to demonstrate, the answer to the latter has to be a clear 'yes'.

Critical responses to donor-driven development: bottom-up development and the post-development critique

The shift towards human development and poverty reduction, as evidenced in the MDGs, suggests an ethicalization of the development process, one that places the human being at the centre. While economic growth has the potential to improve people's lives, concerns that it was not happening quickly enough, that it was not reaching the poorest or that it was directed in ways that served the interests of northern economies have stimulated a range of important critical responses. Three examples of these and their relationships with religions will be explored in Chapters 5–7: human rights and development; gender and development; and environmentally sustainable development. These critical responses have been shaped by approaches to development that consider it as a participatory and 'bottom-up' process rather than a 'top-down' one. From the mid-1970s onwards, approaches to development that aimed at 'putting the last first' (Chambers 1996, 2004) by including people in making decisions about what sort of development they needed and wanted began to emerge. This change opened up space for conceptions of development that differed from the mainstream and that could, among other things, be influenced by people's religions and culture.

Whereas proponents of bottom-up development typically retain faith in a development process, supported by aid and assistance from the Global North, a post-development view takes the critique further. In essence it comprises a radical rejection of the aid business and, as Sumner and Tribe write, it emerged 'as a reaction to the deliberate efforts at progress made in the name of development since World War II . . . triggered in particular by the 1949 Declaration by the US President Truman' (2008: 14). In post-development discourse, the rejection of 'development' does not reject the possibility of positive change and progress, but problematizes the conception and trajectory of development as we know it (Escobar 1995; Kothari 2005; Sachs 1992; Sidaway 2008). As Haynes writes:

> Western interpretations of development represent, understand and talk about the South through a particular prism or lens, called 'development'. It shapes both knowledge and understanding of people in the developing countries; they are as a result passive recipients who uniformly require more and better Western-style 'development' in order to improve the quality of their lives. Post-development discourse analysis finds the developing world uniformly represented as backwards, helpless, problematic, needy and poor.
>
> (2008: 168)

While in reality the hegemony of the donor-driven development project has always been incomplete (where the development aims pursued by individual countries are not necessarily to achieve 'western-style prosperity'), the aid business has tended to proceed as though 'western-style prosperity' is precisely what different countries are aiming for, and assumes that this is an inevitable trajectory. For post-development critics, in a postmodern and postcolonial world, the supremacy of western-style development is ever less certain and can be demonstrated to have exacerbated inequality and poverty. 'Development' is viewed as an elaborate 'hoax', which has served to perpetuate the dominance and interests of the Global North at the expense of the South (Thomas 2000: 19–22).

Among the variety of alternative voices being aired, religious actors in both the Global North and the Global South are an increasingly prominent element of what we might call 'post-development discourse'. These voices are not only articulating opposition towards, and alternatives to, the models pursued within the international aid business; many are also rallying against the secular state-led developmental models pursued in their own countries. Box 2.8 outlines an example of this, the Sarvodaya Shramadana Movement, which promotes a more traditional and small-scale economic and social system based on Buddhist and Gandhian principles as an alternative not only to the style of development promoted by the global aid business, but also to models adopted by the state in Sri Lanka.

Another example of religiously underpinned 'post-development' is found in Rudnyckyj's study of one of the largest state-owned steel enterprises in Indonesia (Krakatau Steel), where the workers had adopted a highly popular human resources programme developed by Ary Ginanjar called 'Emotional and Spiritual Quotient' (ESQ). Rudnyckyj's study investigates how 'the ethical dispositions rooted in religion are cultivated to enable the transition away from state-led modernization' (2011: 3). However, in contrast to Sarvodaya, the challenge posed to state-led modernization is not a revival of traditional, small-scale socio-economic systems, but instead employs spiritual virtues to meld with neo-liberalism for economic success in a globalizing world. The ESQ programme involves 'spiritual training' that combines 'business leadership, human resources, and life-coaching techniques with Islamic practice' (2011: 1) and is promoted as a means of addressing the failure of the large-scale state-

---

Box 2.8

The Sarvodaya Shramadana Movement

Zadek discusses the Sarvodaya Shramadana Movement in Sri Lanka as a 'community organization working to improve the situation of people in rural areas throughout Sri Lanka' (1993: 436). Founded by A. T. Ariyaratne, in 1958, the Sarvodaya Shramadana Movement is an organization based upon Buddhist and Gandhian principles that provides development projects for Sri Lankan villages (Zadek and Szabo 1993; Ariyaratne 1980, 1982; Ariyaratne and Macy 1992). It was the largest indigenous organization working in reconstruction following the 2004 tsunami. As Zadek tells us:

> [T]he critical feature of Sarvodaya's method is embodied in its approach to village-level consultation and mobilization. Sarvodaya has evolved a process of decision-making at village level which would in secular 'development vocabulary' be called participative decision-making (Chambers, 1992). This includes, for example, meetings of the entire village in family gatherings, and the formation of groups within the village (women, youth, elders, etc.), who are then encouraged to articulate their own needs and the path by which those needs might be achieved (usually with some technical, organizational or material help from Sarvodaya).
>
> (1993: 436)

Source: adapted from Tomalin (2007).

led developmentalism of the Suharto regime. State-led development not only has failed to bring about the promised transformations in Indonesia, as in much of the developing world, but in this example is viewed as having failed because of its lack of focus on the cultivation of spiritual virtues. As Rudnyckyj continues, 'Indonesians enmeshed in the resulting spiritual economy did not become just like Westerners, but neither did they remain how they were beforehand' (2011: 22). While conventional 'post-development' discourse typically rejects neo-liberalism and top-down state-led development supported by the global aid business for being tools of the Global North to exploit the South to its advantage, in this Indonesian example the 'post-development' world embraces neo-liberalism. This example reveals a tension within the conventional 'post-development' approach, where in rejecting top-down and centralized development it actually appears instead to be colluding with neo-liberal ideals (Kiely 1999).

* * *

The above discussion has charted the shifts that have taken place in development policy and practice within the agencies and organizations that have shaped the donor-driven development process since the late 1940s, and has examined the implications of these for the role of religion in mainstream development policy and practice. Development studies as an academic discipline has played an important role in influencing these shifts, yet it also has tended to marginalize and avoid the topic of religion. The final section of this chapter will explore the ways in which development studies has traditionally viewed religion and the extent to which it is engaging with the new agenda concerning 'religions and development'.

Development studies and religion: disciplinary responses to a new agenda

Development studies emerged in the 1960s and 1970s with the aim of understanding the causes of poverty and inequality, and identifying theories and practical solutions that can influence the policy of governments and organizations. It emerged as a radical discipline influenced by the revolutionary left-wing politics of the time and in response to the donor-driven 'development' agenda that emerged in the post-Second World War period (Sumner and Tribe 2008: 33–35).

The question of whether or not development studies is a distinct discipline has been debated since its inception. Sumner and Tribe argue that it is 'problem-oriented' rather than 'discipline-oriented' (2008: 66) and that it comprises a combination of disciplines or a 'knowledge community', including, among others, the traditionally secularist disciplines of economics, sociology, political science and gender studies. Over time it has shifted from a 'positivistic' to a more relativistic or 'interpretivist' approach. The former is based on a view that the social world is governed by identifiable laws, much like the physical world. The positivist approach to development studies considers that there are development problems to which definitive solutions can be found, and has been dominated by economists. In contrast, the more relativistic or interpretive approach considers that how development is defined and understood, as well as solutions to development problems, are contextual and varied. This is reflected in the emergence of the post-development critique and participatory approaches to research in development studies, and signals a decline in the dominance of economics (Box 2.9) within the development studies 'knowledge community' and a growth in contributions from other disciplines. Boxes 2.9–2.12 respectively outline some of the key dimensions of the ways in which the disciplines of economics, sociology, political science and anthropology have viewed religion.

Box 2.9

Economics and religion

Jackson and Fleischer write that 'historically, religion has been one of the areas "assumed away" by most economists. . . . The literature reflects a lack of acknowledgement of religion as anything but an obstacle to economic growth' (2007: 1). They identify three types of literature on religion and economics. First, there is that which applies economic theories, including rational choice, risk, game theory or individual utility preferences, to explain patterns of religious behaviour (Anderson 1988; Iannaccone 1998). Second, there is literature that studies the economic consequences of religion, having a specific focus on the arguments proposed by Max Weber (Barro and McCleary 2001, 2003). Third, there are studies looking at 'religious economics', for instance the distinct economic systems that develop in line with religious teachings (e.g. Islamic economics; Zaman 2008). In contrast to some other disciplines (e.g. sociology), economics has tended to have very little to say about the 'substance of religion itself, taking the demand for religion as a given and defining the characteristics of religious communities very loosely' (Jackson and Fleischer 2007: 25). It has also 'largely ignored religion as an influence over behaviour' (ibid.: 26).

Source: adapted from Jackson and Fleischer (2007).

Box 2.10

Sociology and religion

The 'founding fathers' of sociology in the nineteenth century, such as Marx, Durkheim and Weber, gave considerable prominence to religion in their bid to understand the massive social changes that were taking place in an industrializing and modernizing Europe (Tomalin 2007). Today the sociology of religion lies towards the fringes of the discipline, under the influence of views that religion would disappear or become less important as time went on. Not surprisingly, the main focus in the sociology of religion has been upon theories of 'secularization', understood as the process by which religion loses its public role and may eventually even lose its grip on the lives of individuals (Casanova 1994; Woodhead and Heelas 2000).

However, critiques that theories of secularization reflect a western context have since been joined by a broader recognition that religion globally is undergoing a 'resurgence'. In terms of research that has a relevance to the study of religions and development, sociologists have carried out work in a number of relevant areas. For instance, they have carried out research on the role of religions in civil society and social movements (Beckford 2000; Oommen 2004), have looked at the effects of globalization on religion (Beyer 1994, 2006) and have developed theories about social capital and applied these to religion (Putnam 1995; Candland 2000; Berger and Hefner 1998).

Sociologists have been criticized (e.g. by religious studies scholars) for explaining religion in terms of social causes rather than dealing with it on its own terms, thereby potentially missing important dimensions of what is distinctive about religious experience (Wood 1999). Others have been critical that discussions about social capital and religion, for instance, often end up being instrumental since they emphasize the ways in which religion or spirituality can serve social ends.

Source: adapted from Tomalin (2007).

Box 2.11

Political science and religion

Singh *et al.* write that 'of all the disciplines, political science (with perhaps the exception of economics) is the most secular in its outlook' (2007: 3). While '[t]here has, in recent years, been a vigorous interest in political science in examining the relationship between politics and religion . . . very little of the current output is focused on developing countries' (ibid.: 1). The main focus of this research has been on the United States or countries in which it has strategic interests. A review of the *International Political Science Abstracts* from 1990 to 2005 revealed that there were very few entries under religion in the 1990s, and while there has been an increase since the late 1990s, these relate mainly to conflict or causes of conflict, and there are very few references to religions and development.

Where sociologists have been interested in secularization, political scientists have been concerned with secularism, understood in two ways. First, 'it refers to the emergence of an "ideal" secular state, typically, as in the USA, characterised by the creation of a "wall of separation" between "church" and state. The exact form this separation has taken has varied considerably in different states, resulting in varieties of state secularism (e.g. USA, France, India)' (Singh *et al.* 2007: 3). Second, 'secularism is associated with the broader process of the secularisation of society, in which the influence of religion and religious institutions gradually declines' (ibid.: 3). Despite the fact that the categories of political science do not translate well in developing countries – for instance, notions of distinct 'churches' that have been separated from the state through particular historical processes are a largely western phenomenon – this understanding of the separation of religion and the state has impacted upon the politics of developing societies since 1945, and 'for most emerging post-colonial leaders, secularism was a *sine qua non* of development and democracy . . . with the aim of emulating the social and political transformation in the West' (ibid.: 3–4). By the 1980s, in the developing world the secular state was in retreat, reflecting the return or resurgence of religion (Thomas 2005). This has influenced both global politics and political science, where we find literature on a number of themes, from the role of religion in democratization and good governance, to its impact on human rights and social capital, as well as the complex links between ethnicity, religion and conflict.

Source: adapted from Singh *et al.* (2007).

The impact of economics in particular has had an influence on the marginalization of religion within development studies, with its view that economic growth and religiosity were mostly incompatible (Box 2.9). However, theories of secularization, honed within sociology in the West, were accepted without questioning the extent to which they were likely to be valid for developing contexts (Box 2.10), as was the emphasis on secularism within political science (Box 2.11). Ver Beek carried out a content analysis of the three leading development studies journals between 1982 and 1998 (see Table 2.1), finding

> only scant reference to the topics of spirituality or religion. In fact, two of these journals contain not one article during this period in which the relationship between development and religion or spirituality was the central theme. A policy review of three influential development organisations also demonstrated not only that these have no policy on how to treat the area of spirituality but that they consciously seek to avoid the topic in their programmes.
>
> (2000: 60)

One notable break with this pattern was the publication in 1980 of a special issue of the journal *World Development* on the topic of religion. The authors of the introductory article, Wilber and Jameson, argue that the model of

Table 2.1 *Number of articles with references to the listed keywords, by journal*

Journal, 1982–1988	Keyword				
	Environment	Gender	Population	Spiritual and spirituality	Religion and religious
World Development	83	85	89	0	5
Journal of Development Studies	19	46	38	0	1
Journal of Developing Areas	18	32	43	0	10

Source: Ver Beek (2000: 37).

development as economic growth was only able to deal with religion in a very limited sense: first, in instrumental terms with respect to how it can influence GDP; second, as a social institution that is considered to be an impediment to development; third, as a private matter and having no public influence, which frees energies for the individual to engage in economic activity; and finally, as a phenomenon that will eventually be undercut by successful development (1980: 471). They consider that these views about religion reflect a western context and that when these assumptions about religion, as well as the benefits of rapid economic development, are transported to developing countries, they are likely to fail and produce greater inequality.

However, in this landmark article Wilber and Jameson argue that the shift from a focus on development as economic growth in the 1970s, and the rise of 'needs based approaches' (religion, they suggest, should be considered as a 'need'), mean that 'it is appropriate to reconsider the relationship between religious values and development' (1980: 468). This view is further reinforced, they argue, by the emergence of 'growth-with-equity' models, which in themselves do 'not discount the possibility and even desirability of growth but which suggest limits on that process that make it qualitatively different' (ibid.: 475–476). In particular, they suggest that in thinking about how to achieve appropriate development in different contexts, in a way that meets basic needs and encourages both growth and equity, the role that religion plays 'in defining development and putting limits on acceptable definitions' (ibid.: 468) should be considered. Moreover, they argue that unless approaches to development are consistent with '*the inherited moral base of society*' (ibid.: 468; emphasis added), which is shaped by religion, they are likely to be ineffective. For them, the 'moral base of society' comprises the collectively agreed-upon value system which means that a society can function and reproduce itself over time, and in most developing countries, they argue, this continues to be shaped by religion.

They do not view the 'moral base of society' as something that is rigid and unchanging, but are concerned that the rapid economic growth pursued by donor-driven development since the Second World War had not allowed sufficient time for the moral bases of developing countries to adapt. Instead, 'the onslaught of rapid growth and the spread of individualism' (1980: 474) meant that, while the entrepreneurial elite was freed from religious restrictions, it became liable to corruption and dishonesty, since the new economic order did not provide the basis for social ethics. However, the non-elite, who felt that their religious values had been challenged by these rapid changes, were more likely to rise up to protect and promote those values in the face of the inequalities exacerbated by development policies and practices. In order to build a development process that reflects a society's moral base, Wilber and Jameson argue for a research agenda in development studies that does four things: (1) looks at the ways in which religion shapes individual values; (2) investigates the role that religion might play in blocking development efforts; (3) researches religion as a positive impulse in development; and (4) looks at the growing role of religions as transnational actors.

This article was in many ways ahead of its time and was unable to generate much support for its analysis and suggestions. While these ideas are now being explored in a more varied and committed way in development research, Deneulin and Bano write that 'the original contribution of the 1980 World Development – its argument that religion ought to be considered as the moral basis infusing the very concept of development – has still to permeate today's development thinking and practice' (2009: 40).

Box 2.12

Anthropology: culture, religions and development

Apart from theology and religious studies, the study of religion has had its greatest prominence in anthropology, which acknowledges the 'importance of religion in understanding everyday life and human behaviour' and that it is 'one important foundation of cultural beliefs, identities and actions/practices' (Bradley 2007: 7). Anthropology is essentially concerned with the study of human behaviour, and 'anthropologists are concerned to understand how people perceive their identity and the world around them. They also document processes of social change' (ibid.: 1). Whereas for an economist the nature of religion is not important for understanding its impact or the reasons why people choose to participate in it, for an anthropologist this detail is of far greater significance. Religion is typically seen by anthropologists to be a part of the broader category of 'culture'. In contrast to Tylor's broad definition of culture given in Box 1.1 on p. 2, a more specific and narrower definition, focusing on the cognitive dimensions of culture, is provided by Geertz: 'an historically transmitted pattern of meanings embodied in symbols, a system of inherited conceptions expressed in symbolic forms by means of which men [*sic*] communicate, perpetuate, and develop their knowledge about and attitudes toward life' (1973: 89).

Religions as forms of culture comprise different sorts of beliefs and practices that are oriented towards understandings of the 'sacred'. These include practices outside the mainstream religions, such as shamanism and witchcraft, as well as the rituals, myths and doctrines of the so-called world religions. One of the reasons why anthropology has focused on the study of religion is because it originated in a bid to better understand the culture and society of the European colonies, where religion was perceived to be of primary importance in people's lives. This has, however, led to critique of anthropology as a colonialist discipline. Clifford and Marcus's influential book *Writing Culture* (1986) 'was founded on the acknowledgment that traditional anthropology produced holistic representations that were fundamentally the products of western exoticised perceptions of the lives of eastern peoples . . . [and it] forced anthropologists to address the question of who, what, how and why might we represent?' (Bradley 2007: 3).

Anthropology has influenced development studies through demonstrating the benefits of micro-level, ethnographic qualitative research that aims to understand how people perceive their lives and their interactions with others (e.g. the World Bank's *Voices of the Poor* studies (e.g. Narayan *et al.* 2000a, 2000b)). A focus on anthropology and development developed during the 1990s, and three types of literature can be identified:

first, anthropological studies of the development process, which are often critical of the ethnocentric bias of development policy and practice (Grillo and Stirrat 1997); second, the construction of ethnographic techniques that can be used as part of the development process (Chambers 1992, 1996, 2005); and third, anthropological studies of 'development' at the grassroots level (Bradley 2007: 48). While a substantial body of literature that directly addresses the relationships between religions and the formal development process from an anthropological perspective is still to evolve, much anthropology of religion is in effect concerned with the influence of religion on development as social change.

Source: adapted from Bradley (2007).

Despite the emphasis upon 'positivistic' and economistic conceptions of development, as Box 2.12 suggests, anthropology in particular has greatly influenced the emergence of a more 'interpretivist' approach. The groundbreaking World Bank *Voices of the Poor* studies (e.g. Narayan *et al*. 2000a, 2000b), which broke with orthodoxy in approaching the issue of poverty through asking what it meant to people rather than through collecting data and producing statistics, reflects this influence. It was initially instigated by the World Bank in preparation for its *World Development Report 2000/2001* and resulted in three volumes. The first volume involved discussions with over 40,000 poor women and men in over fifty countries in the 1990s, using Participatory Poverty Assessment (PPA), which makes use of 'unstructured interviews, discussion groups, and a variety of participatory visual methods' (Narayan *et al*. 2000a: 16). PPA captures information that standard poverty assessments miss, since they tend to focus on the collection of quantitative data derived from large-scale surveys about factors such as household income, asset ownership, consumption patterns and access to services, as well as social and demographic profiles of the poor (ibid.: 15). By contrast, PPA methods use open-ended questions allowing 'the emergence of issues and dimensions of poverty that are important to the community but not necessarily known to the researchers' (ibid.: 16). Moreover, while traditional household surveys take the household as the primary unit of assessment, PPA deals with men and women separately, and treats them as having distinct interests and needs.

One of the important findings of the book was that religion is crucial to people's sense of well-being: 'the poor often mention turning to God for comfort, solace, and support' (Narayan *et al*. 2000a: 38). 'Cultural identity', which is shaped and maintained through a range of activities, including religious customs, rituals and festivals, was also found to be important to people: 'societal bonds can help stabilize communities and ease the psychological stresses of poverty' (ibid.: 42). The second volume in the series (based on research in 1999 with over 20,000 people in twenty-three countries) characterizes poverty in different contexts as associated with 'powerlessness' and points out that 'faith-based organizations' (here taken to mean places of worship rather than 'faith-based' development organizations) can be both disempowering and

Box 2.13

Voices of the Poor

Voices of the Poor presents a challenging conclusion, one that has dramatic consequences for development practice if implemented. Considering religion as an intrinsic component of people's well-being, alongside health, education, shelter, material security and others, transforms conventional development practices which have so far ignored the religious dimensions.

(Deneulin and Bano 2009: 44)

The poor person has to exist so he can serve the great one, the rich. God made things like that.

(Quotation from an informant in Brazil, 1995: Narayan *et al.* 2000a: 37)

As God gives food to a tiny insect living in the stones, He makes sure we have enough food to live.

(Pakistan, 1996: ibid.: 38)

The mosque is our court, school, and lawyer, while the village council is of no support, and policemen just provoke disorder. If something is stolen, the police do not search for the thieves, and, even if the thieves are caught, no further action is taken.

(Urmaral, Kyrgyz Republic: Narayan *et al.* 2000b: 224)

There is a huge tree at the bank of one of the seven crater lakes in the area. People go to the tree on a Sunday after Meskal [a church festival] with wet straw in their hands. The wet straw symbolizes the desire to have 'wet land,' 'wet hands,' etc. Wet things are supposed to stand for prosperity and wet land allows growth. The main purpose therefore is to pray to God to make the land wet with rain. People said, 'We believe in it and it works; we get together and pray when we need something desperately; we go and pray for our children's health.'

(Community report, Kajima, Ethiopia: ibid.: 222–223)

empowering; their role can vary 'from being a balm for the body and soul to being a divisive force in a community' (Narayan *et al.* 2000b: 222). However, they were frequently cited by people when discussing the most important institutions in their lives. Overall, 'faith-based organizations' were among those institutions most trusted by people, in contrast to state institutions, which were typically considered to be ineffective (ibid.: 199–201). Some extracts from these studies are presented in Box 2.13.

Summing up

The main aim of this chapter was to chart shifts in approaches to donor-led development that have taken place over the past several decades which mean that with respect to religion there has been a shift from 'estrangement' to 'engagement' (Clarke and Jennings 2008a: 2). This evolution has taken place alongside broader social, cultural and political changes globally over the past decades that have thrust religion back onto the public stage. Thomas (2005: 223–224) captures the main points raised in this discussion in terms of five factors, which he considers have contributed to a more holistic understanding of development over the past thirty years, meaning that religion and culture are more likely to enter into international development agendas:

1 There has been a reassessment of the emphasis on economic growth and industrialization within the international aid business.
2 Development actors have begun to recognize the costs of disregarding people's values in the quest for western understandings of development and modernization – 'people want to develop, they want the fruits of economic prosperity, but without losing their souls' (Thomas 2005: 223).
3 There is now less inclination to see development in 'positivistic' and 'value-free' terms, giving rise to critical perspectives that aim to ethicalize the development process as well as those that focus upon 'post-development'.
4 Culture is becoming part of the official development agenda (for instance, in 1992 UNESCO established the World Commission on Culture and Development).
5 Finally, Thomas discusses what he calls the 'larger crisis of modernity', of which the resurgence of religion is part. What he means here is that the assumptions underpinning the modern project, for instance the links between modernization and secularization, have been challenged. Not only have the desirability and inevitability of western approaches to modernization become increasingly uncertain, but so too have predictions about the future of religion.

In the following chapter we will look in more detail at theories about modernization and secularization, and their relevance for thinking about the relationships between religions and development. In particular, I will argue that there is a need to question the extent to which the familiar definitions, concepts and theories that we typically have at our disposal for the study of religion are adequate for an examination of 'religions and development'. This is because what is typically meant by religion and religious change in the West is often poorly suited to approaching these phenomena in developing contexts.

Questions for discussion

1 What is 'development'?

2 How have approaches to development evolved since the 1950s, making it more likely that religion and culture will be taken seriously by development actors?

3 How have the different academic disciplines that constitute the development studies 'knowledge community' dealt with the topic of religion?

Recommended reading

Desai, Vandana and Rob Potter (eds) (2008) *The Companion to Development Studies*, **2nd edn, London: Hodder Education.** This is one of a number of useful edited volumes providing an overview of key themes and issues within development studies.

Special Issue of *World Development* **on Religious Values and Development (1980, 7–8).** This ground-breaking volume includes fourteen papers on a range of topics, from 'The Role of Buddhist Monks in Development' and 'African Independent Churches and Economic Development' to 'Islam and Development: The Zia Regime in Pakistan' and 'Attitudes toward Development among Catholics in Sri Lanka'.

Thomas, Scott M. (2005) *The Global Resurgence of Religion and the Transformation of International Relations: The Struggle for the Soul of the Twenty-first Century*, **Basingstoke, UK: Palgrave Macmillan.** Thomas is interested in the current global rise of religion as comprising more than a 'clash of civilizations' underpinned by religious fundamentalism. He devotes one chapter to an examination of religions and development.

Websites

http://religionsforpeace.org/resources/toolkits/faith-in-action.html Religions for Peace, a supporter of the UN Millennium Campaign, has developed a toolkit to engage and equip religious leaders and their communities in the fight against poverty and disease.

http://www.religion-and-development.nl The Knowledge Centre Religion and Development website provides a range of papers, reports and links to other websites, concerning the relations between religion and development. It has been set up to assist with Dutch development efforts but has relevance beyond this focus.

③ Concepts and theories for studying religions globally

This chapter will:

- discuss the difficulties involved in defining religion;
- evaluate different disciplinary approaches to the study of religion;
- examine the ways in which the category 'religion', and in particular the idea of so-called world religions, has been contested as a western Christian-influenced and modern concept that has come to shape how religions are viewed globally;
- discuss the terms 'secularization', 'secular' and 'secularism', with a particular focus on the postcolonial deconstruction of these terms;
- summarize the main challenges raised by this chapter for understanding the relationships between religions and mainstream development studies, policy and practice.

Introduction: religion as a contested category

> All too often theorists have taken religion as a relatively unproblematic unitary and homogeneous phenomenon that can be analysed and compared across time and space without proper consideration of its multi-faceted and socially constructed character.
>
> (Beckford 2003: 15)

While mainstream development research, policy and practice have tended to ignore religion as a relevant factor, Thomas writes that 'it is now more widely recognized that successful development, no matter how it is defined, can only occur if social and economic change correspond with the moral base of society' (2005: 222). However, the discussion in Chapters 1 and 2 treated the category of religion as though it is more or less self-evident and unproblematic. It was assumed that there is a cross-cultural phenomenon called 'religion' (often viewed in terms of the distinct so-called world religions of Christianity, Hinduism, Buddhism, Islam, etc.) that is found in all societies in one form or another. It was also taken for granted that this thing called 'religion' can be defined in such a way as to make it possible to identify what is religious and what is not. After all, without these assumptions being made, the topic of this book – 'religions and development' – would be meaningless.

However, as early as the 1960s, scholars began to question the extent to which the category of religion, as well as the distinction between the religious and the non-religious, can be treated in such a straightforward manner. Instead, they have put forward the view that the 'world religions paradigm' in the study of religion actually reinforces western colonialist discourses about what a religion should look like. Moreover, they argue that the study of religion, as well as our understanding of what a religion is, involves transposing a western Christian model of the distinction between the religious and the secular (which emerged following the sixteenth-century Protestant Reformation in Northern Europe) onto non-Christian and non-western contexts, thereby distorting our understanding of those 'religions' as well as transforming the way in which they come to understand themselves (Smith 1991 [1962]; Asad 1993, 2003; Beyer 2006; Fitzgerald 2000; King 1999).

This chapter explores these arguments and assesses the implications of the following questions for thinking about the relationships between religions and international development:

- If we are employing a constructed term and associated concepts (e.g. 'secular' or 'secularism') to think about the relevance of what we are calling 'religions' to international development processes and outcomes, then are we in danger of failing to understand the relationships that we wish to probe and illuminate?
- In what ways do these western-influenced understandings of the nature of religion shape approaches to predicting its future (e.g. theories of secularization)?
- How do we accommodate these critiques alongside the fact that these terms and concepts are widely used and understood across the globe: they are not just employed by westerners attempting to describe certain types of human behaviour and value systems encountered elsewhere?

In this book I suggest that it is preferable to use the plural *religions* (e.g. '*religions* and development') to draw the emphasis away from the assumption that there is a universal and ahistorical thing called religion that can be discovered by scholars interested in understanding its impacts and interactions with other sorts of social and political processes, including international development policy and practice. Instead, there are a variety of 'religions' that do not neatly fit universal definitions, and the categories and descriptions that we have become accustomed to using are socially constructed. Moreover, particular 'religions' are themselves internally diverse and there can be multiple versions, each claiming that they represent the legitimate tradition.

We will begin this critical discussion about the term 'religion' and its associated concepts with an examination of attempts to arrive at a universal definition of religion. Rather than suggesting that we need to work harder to find a definition, the failure to successfully achieve this instead forces us to question whether or not the term and concept 'religion' is a useful cross-cultural tool.

Defining religion

There is no agreed-upon definition of what (a) religion is and there have been a range of ways of approaching this question from different disciplinary perspectives (these disciplinary approaches to the study of religion will be discussed in more detail below, p. 57ff.). Thus, while there is overlap, a theologian's definition is likely to differ from that of a sociologist, and an anthropologist will again take a different view. Moreover, one's personal religious belief may also affect how (a) religion is defined and understood. As Connolly writes, just as for social phenomena,

> [a]ccurate, objective accounts of religious phenomena and religious traditions simply do not exist in their own right. All accounts of religion are accounts by people who approach their study from a particular starting-point. They bring with them a set of assumptions about the nature of their subject-matter and, however naïve it might be, a method or set of methods for obtaining information and making sense of it.
>
> (1999: 1)

Despite the difficulties involved, scholars have argued that it is necessary to attempt to define religion in order that they can identify it as an entity to be studied, or so that they can determine its influence. However, these days there is more awareness that such definitions may only be a starting point and that we need to be flexible in how we think about what religion is (Lincoln 2003). A broad distinction can be drawn between *substantive* and *functionalist* definitions of religion, although some definitions combine both elements (Furseth and Repstad 2006).

Substantive definitions of religion

Substantive definitions concentrate on *what a religion is*. This would include the apparent cross-cultural attributes of religion that distinguish it from other social phenomena, particularly belief in a transcendental reality and/or (a) spiritual being(s), religiosity (which is signified by the beliefs held and practices in which adherents engage) and affiliation to a religious organization. As Haynes writes, religion can be viewed as 'a system of language and practice that organizes the world in terms of what is deemed holy; and with ultimacy, that is it relates people to ultimate conditions of existence' (2011: 13). Box 3.1 outlines the definition of religion provided by the religious studies scholar Frank Whaling.

Box 3.1

Whaling's definition of religion

Frank Whaling's model consists of eight components apparently found in all religions: religious community, ritual and worship, ethics, social and political involvement, scripture/myth, concepts, aesthetics and spirituality (1986: 38ff.). Beyond these tangible aspects he identifies a non-observable 'transcendent reality' in each religion: [f]or a Christian, this is God; for a Muslim Allah; for a Jew Yahweh; for a Hindu Brahman, and for a Buddhist Nirvana (ibid.: 46). Likewise all religions have a 'mediating focus', which provides a link to the 'transcendent reality': 'Christ for a Christian, the Koran for a Muslim, the Torah for a Jew, a personal deity or the Atman for a Hindu, and the Buddha or the Dharma for a Buddhist' (ibid.: 46).

While Box 3.1 suggests that the naming of and understanding of 'transcendent reality' differs between religious traditions, it is also the case that this may differ even within religious traditions. Moreover, questions about the exact nature of divine beings in religions have caused disagreement between members of different academic disciplines. Whereas religious believers and theologians (who tend to be religious believers themselves) accept the actual existence of divine beings, other academic disciplines typically reduce belief in transcendent reality to 'material' causes (e.g. social, psychological and even physical).

Substantive definitions of religion have been critiqued for being guilty of deciding before an investigation what a religion is or should be (particularly if they aim to be very detailed and comprehensive, rather than broad or partial). This supposed failing is not unique to attempts to define what a religion is, however, since any approach to defining shapes our understanding and reveals certain assumptions about the nature of the phenomena being defined. A reliance on substantive definitions of religion can, therefore, mean that the researcher fails to investigate all actual forms of 'religiosity'. Even Edward Tylor's minimal definition of religion (1871) as 'belief in spiritual beings' would mean that Buddhism (which does not teach the existence of a supreme divine entity) is not counted as a religion, although in important senses it resembles what we normally think of as a religious tradition.

Moreover, empirical studies of religion demonstrate huge variety not only in beliefs and styles of practice, but also in the 'forms' that religions take. A distinction is sometimes drawn between those religions that focus on 'orthodoxy' (right belief) and those that focus on 'orthopraxy' (right action). Examples of the former include Christianity, for instance. Within the latter type are located Asian traditions such as Hinduism or Confucianism, where there is more emphasis upon correct religious practice and one's social role vis-à-vis the conventions of that religion than upon believing in particular doctrines or teachings. Thus, definitions of religion that emphasize belief and doctrine may be unable to capture the relative unimportance of these within some traditions outside the Abrahamic religions (i.e. Christianity, Islam and Judaism).

Critics argue that substantive definitions of religion cannot provide definitions of what religions really are. This is not only because religions are so diverse that it is virtually impossible to draw out any meaningful cross-cultural patterns, but also because such definitions are only capable of reflecting what scholars of the study of religion have conventionally understood religion to be, reflecting a western colonial and Christian heritage. However, according to some critics, attempting to define religion not only assumes that religions have identifiable cross-cultural attributes, which actually reflect western Christian ideas of what religion should be, but also that the very category of religion, and more specifically of distinct 'world religions', is itself a western invention that emerged following the Protestant Reformation in the sixteenth century (Asad 1993, 2003; Beyer 2006; Fitzgerald 2000; this is discussed later in the chapter). In Box 3.6 I apply this critique to the question of whether or not Hinduism is a 'world religion': prior to the eighteenth century, the word 'Hinduism' was unknown in the Indian subcontinent, only appearing in response to the coining of the term by British colonialists.

Functionalist definitions of religion

In contrast to substantive definitions, functionalist definitions are primarily concerned with *what a religion does*: the role that it plays in both the construction of people's worldviews and the maintenance of social cohesion, although scholars differ in their opinion about the relative importance of these. Functionalist definitions have been particularly popular within sociology. Whether religion is considered to have a positive or a negative function is open to investigation and may be influenced by the views of the researcher.

As was demonstrated in Box 2.4 on p. 29, for Karl Marx religion played a crucial role but as a response to the social alienation inherent in the capitalist system rather than as a positive social force. This view is expressed in his now famous depiction of religion as the 'opium of the people'. Emile Durkheim was similarly interested in the function that religion performed and also drew attention to its role in maintaining social institutions. In his book *Elementary Forms of Religious Life* (1995 [1912]) he provides a study of Australian Aborigines and argues that their totemic gods are expressions of their understanding of society. Thus, for Durkheim religion is an expression of society upon which humans impose notions of the supernatural: religion has social rather than 'divine' causes, and when we worship gods, we are actually worshipping society. This, he maintains, is true of religion in all societies.

Some contemporary studies of Durkheim's sociology of religion have applied his thinking to considerations of the role of phenomena such as sport, as a style of 'modern religion' that performs similar functions to traditional religiosity. This points to a more general critique of functionalist definitions of religion as

Box 3.2

Clifford Geertz's definition of religion

Geertz (1973: 90) defined religion as '(1) a system of symbols which acts to (2) establish powerful, pervasive, and long-lasting moods and motivations in men [and women] by (3) formulating conceptions of a general order of existence and (4) clothing these conceptions with such an aura of factuality that (5) the moods and motivations seem uniquely realistic'.

being potentially too inclusive and therefore losing sight of what, many would argue, separates the religious from the profane or secular. While the functionalist definition of the anthropologist Clifford Geertz outlined in Box 3.2 has been one of the most well used, as Glazier writes, it 'may be useful in elaborating what religion is like conceptually and what it does psychologically and socially, [it] has been criticized for failing to explain specifically how a researcher might identify religion when encountered in the field' (1998).

Should we attempt to define religion?

Definitions of religion have, therefore, been viewed as either too general or too specific, presupposing not only what a religion is like but also that religions exist in the way they are imagined to by (western) scholars of religion. By contrast, as Box 3.3 suggests, the sociologist Max Weber refused to define religion. Others have argued that he must have had an implicit substantive definition of religion (although this is never clearly outlined in his work) since he 'included some things and excluded others' in his research and writing about religion (Aldridge 2007: 30). However, this circularity affects anyone wishing to research 'religion' but not wanting to impose an understanding of it at the outset: if you do not have some sort of understanding of what a religion is, then how can you set out to study it?

Box 3.3

Max Weber's reluctance to define religion

Weber (1963 [1922]: 1) wrote, '[t]o define "religion" to say what it is, is not possible at the start of a presentation such as this. Definition can be attempted, if at all, only at the conclusion of the study. The essence of religion is not even our concern, as we make it our task to study the conditions and effects of a particular type of social behavior. The external courses of religious behaviors are so diverse that an understanding of this behavior can only be achieved from the viewpoint of the subjective experiences, ideas, and purposes of the individuals concerned – in short, from the viewpoint of the religious behavior's "meaning"'.

The real import of Weber's statement is his reluctance to posit any definition of religion that imposed an 'outsider' understanding upon the meaning of religious behaviour and experience for the individual. In particular, this refers to his view that the researcher should not comment on the nature of the truth claims made by religious adherents. He saw religion as separate to or autonomous from society (that is, it is not a product of class interests or of the collective social consciousness) and hence separated the issue of the truth claims of religious traditions from his analysis, thereby avoiding the reductionism of Marx and Durkheim. This 'bracketing of truth claims' came to influence the 'methodological agnosticism' found in the phenomenological approach that has shaped religious studies as a discipline (see what follows).

In contrast to Marx and Durkheim, he placed his emphasis upon religion as a system of meaning for the individual rather than upon its social function. Moreover, as Dillon writes, his writings on religion 'demonstrated both the significance of different historical and cultural contexts on the evolution, development and societal implications of different religions as well as drawing attention to the intricate cultural intertwining of religion and societal structures' (2009: 410). His work drew attention away from any assumptions that there are true or universal versions of religious traditions, effectively immune to social influence, and instead highlighted that religion itself is constructed by and constructs social forces, and is thus highly varied in its meanings and manifestations.

Does this mean that we should abandon attempts to provide a definition of religion? Or should we adopt different definitions depending upon the context? Is the concept 'religion' a useful cross-cultural category or is it simply too likely to be influenced by particular understandings such that it becomes meaningless when applied to other contexts? Like Weber, should we 'bracket the truth claims' of different religions as beyond academic scrutiny, or can they be explained in terms of social factors?

These are questions that the academic study of religion has grappled with since the eighteenth century, yet they still have great saliency today. But what is their relevance for our interest here in the study of religions and development? One of the dangers of adopting an approach to studying religion that assumes that it has fixed cross-cultural attributes is that religions are diverse and not all features of, for instance, popular substantive definitions are found across contexts and religious traditions. As has already been discussed, a view of religion that focuses on the importance of texts, teachings and belief is not universal, in the past or today. An assumption that these things are important, however, can lead to an emphasis upon attempting to identify what particular texts or teachings in different religions tell us about key development issues, rather than examining how people live out their religion in practice. Thus, there is a need to be more nuanced and contextual in how we understand what a religion is in different contexts. This suggests (following Weber) the benefits of an approach in which one has an intuitive and provisional understanding of what a religion is and where to find it, thus enabling us to begin our investigation, but a more detailed description and understanding (and possibly a definition, albeit not a universal one) can only be achieved at the end of a study.

This discussion has drawn attention to the problems inherent in attempting to provide definitions of religion. In particular, not only are religions so diverse that it can be difficult to identify a set of cross-cultural attributes, but also definitions of religion are only capable of reflecting what scholars of the study of religion have conventionally understood religion to be, reflecting a western colonial and Christian heritage. Before we look in more detail at the claim that 'religion' is a western colonial and Christian-influenced invention, it will be helpful to examine what different disciplines consider religion to be (i.e. what sort of phenomenon) and what its significance is. The following section examines the ways in which different academic disciplines have approached the study of religion and what these tell us about how religion has been conceptualized in the (western) academy.

Approaches to the study of religion: an historical overview

Insider and outsider approaches

Whether one is a religious adherent or not, as well as one's disciplinary background, can have an impact upon how religion is studied. The study of religion (as opposed to theology) can be located in scholarly developments since the eighteenth century, particularly within anthropology, sociology, psychology and philosophy (for introductory texts, see Fenn 2003; Dillon 2003; Bowie 1999; Banton 1966; Parsons and Jonte-Pace 2001; Davies 2004). Theology, by contrast, is often described as an 'insider' approach, which aims at knowledge of God and his relationship to humanity and the world. Literally, the word 'theology' is derived from the Greek language, where *theos* means 'God' and *logos* means 'word, study, science'. Thus, theology equates to 'the study or science of God'. Different theologies accept the truth claims of the religious tradition they pertain to and are most often carried out by people who are members of that tradition. As Thomas and Wondra write, with respect to Christian theology:

> Theology is normative in attempting to determine the true Christian faith – to distinguish true doctrine from false doctrine, true practice from false practice, true worship from false worship; to determine what is and what is not part of the Christian faith. Thus, one of the functions of theology is to scrutinize, criticize, and, if necessary, reform the church's formulations of its faith in creeds, conciliar decisions, and confessions, and its practices in worship, prayer, mission, and daily life.
>
> (2002: 5)

While the word 'theology' is normally used with respect to the Christian tradition, it can also be used to apply to the 'insider' study of religions more generally. However, different religious traditions may also have their own terms to describe this type of activity. For instance, in Buddhism we are more likely to find the term 'philosophy' used, primarily because Buddhism does not teach the belief in a divine power (and *theos* is the Greek word for God). In Islam an enterprise of equivalent status to theology in the Christian tradition is the study of Islamic law, or *fiqr*.

With the rise of the social sciences, including anthropology (which arose in response to colonial needs), we find the emergence of 'secular' or 'outsider' approaches to the study of religion, which seek to understand how and why humans engage in religious belief and action. Although in practice it is not always easy to differentiate 'theological' or 'insider' from 'outsider' approaches, since many sociologists or anthropologists, for instance, who study religion may belong to the traditions they are investigating, it has proved to be a method-ologically useful distinction and one that prevails today. While there are overlaps between theological and other approaches to the study of religion, the variety of methods, concepts and theories that different disciplines have brought to the study of religion has been hugely important in improving understandings of this diverse and complex phenomena.

However, there are also important differences between theological and social science approaches to the study of religion that can be traced to differences in their ontological (*what is the nature of reality?*) and epistemological (*what can we know?*) underpinnings. For theology, God is real and can be known via different methods, including direct experience, observation of the natural world and reason. As McGrath writes:

> How can God be known? For some, God is to be sought out within the complexities and ambiguities of the world. The 'human quest for God' involves the careful weighing of evidence drawn from the natural world, including human reason and conscience. For others, human nature is limited in its abilities and is unable to discern the existence or nature of God in this way. Humanity needs to be told what God is like. The issue being debated is fundamentally that of revelation.
>
> (2011: 22)

By contrast, for the 'positivist' social scientist the only reality is what we can directly observe (e.g. the social, political, economic, psychological or material worlds), and the existence of God is denied and/or reduced to social or psychological causes. This view of reality and knowledge was influenced by the Scientific Revolution (of the sixteenth and seventeenth centuries) and the Enlightenment (in the seventeenth and eighteenth centuries) in Northern Europe and has underpinned the ontological and epistemological frameworks across the social sciences, including anthropology. Early 'positivism' in sociology, for instance, closely mirrored the scientific approach, where the dominant view was that society was real and the rules that govern it could be discovered through empirical observation. Moreover, in this endeavour the researcher was considered to be objective and neutral, simply uncovering pre-existing universal laws governing social relations through empirical and theoretical research.

During the twentieth century, positivism in the social sciences became challenged by relativist or interpretivist approaches (also sometimes known as postmodernism, constructionism, constructivism and hermeneutics; Sumner and Tribe 2008: 58), which view social or political reality as constructed, acknowledge the subjectivity of the researcher and are interested in uncovering the meaning that people invest in their social lives and actions (rather than universal laws). These opposing social science perspectives also show preferences for different approaches to theory and method, although there is overlap. Positivists favour quantitative methods that use 'top-down' deductive theories to guide them. These involve beginning with a theory, generating a hypothesis, undertaking observation and confirming (or not) the hypothesis. Interpretivists, by contrast, prefer qualitative methods, and 'bottom-up' inductive theory, which begins with observation, identifies possible patterns, proposes a hypothesis and suggests a theory. Thus, we can see that the attempt to put forward universal definitions of religion, as discussed above, is to be located within a positivist social science tradition, whereas the Weberian approach, which is more interested in meaning and local understandings, draws upon interpretivist or relativist ontology and epistemology.

Moreover, in contrast to theology, these 'secular' social science disciplines have tended to understand religion in terms of non-religious functions as well as causes or origins. This could include viewing religion as a way for humans to regulate behaviour via conceptions of right and wrong (rather than as an actual response to the requirements of a divine power) or a view of religion as a pragmatic attempt to control our physical environment and to explain the existence of the material world. Moreover, 'religious phenomena' have been attributed to 'naturalistic' causes (e.g. those that are social or psychological) rather than supernatural ones. A tension, therefore, has arisen between those who are more inclined to adopt naturalistic explanations and those who argue that there can be no other cause of religion than the supernatural.

The emergence of religious studies

Over time, a third option developed, with the emergence of 'religious studies' in the 1960s, which sought to distinguish itself from both the confessional nature of theology and the religious reductionism of social sciences such as sociology and psychology. Religious studies aimed to create space for the 'secular study of religion', through adopting a '"non-confessional approach" which tried to treat religions as key dimensions of human culture which can be understood in ways akin to other discipline's understandings of their objects' (Flood 1999: 18). In taking religion as the disciplinary focus, supporters of this new approach argued that to try to explain religion in terms of the tools of other disciplines – for instance, social or psychological processes – ultimately diminishes our understanding of the significance of religious belief and experience for the individual. Moreover, this band of scholars also objected to the Eurocentric tendencies of anthropological and 'Orientalist' accounts of religion when they implied that such traditions were inferior to, or to be contrasted with, the Christian tradition. The challenge facing religious studies was to develop a methodology that was capable of two things. First, it needed a methodology that enabled the production of scholarly reflection upon all religious traditions on their own terms, but by researchers who stood 'outside' the religious traditions they were studying (unlike theology, where theologians were 'insiders'). Second, it was deemed important that this method should remain agnostic about the truth of religious beliefs, since to reduce them to naturalistic explanations was to take something away from the religious subject's account, and to affirm them would imply a theological position.

To meet this challenge, a phenomenological method had become popular by the 1960s, which demanded that the researcher 'bracket' his or her own interpretation of religious phenomena and instead endeavour to enter into empathy with the believer in order to describe and understand their religious beliefs and actions. As described in Box 3.4, this empathetic position, it was argued, would allow the researcher to describe as closely as possible what the religion means to the believer (the *eidetic vision*). While this phenomenological method does appear to resemble anthropological methods involving close study within different communities in order to understand what people do and why, as Glazier writes, with respect to religious belief, 'the overwhelming majority of anthropologists are skeptics. Most anthropologists are materialists and reductionists. They would find themselves in strong agreement with Firth . . . who contends that "there is truth in every religion. But it is a human not a divine truth"' (1998).

Thus, religious studies sought to distinguish itself from theology – where 'the language of Theology is a language which *expresses* religion whereas the language of Religious Studies is a language *about* religion' (Flood 1999: 20) – but also from invoking alien or foreign categories to explain religious phenomena. As Box 3.5 illustrates, this is also the position adopted by the social constructionist Peter Berger, who argues that a reductionist position, which

---

Box 3.4

The phenomenological method

According to Smart (1996: 2; also 1973, 1978):

> The use of *epoché* or suspension of belief, together with the use of empathy, in entering into the experiences and intentions of religious participants . . . implies that, in describing the way people behave, we do not use, so far as we can avoid them, alien categories to evoke the nature of their acts and to understand those acts. In this sense phenomenology is the attitude of informed empathy. It tries to bring out what religious acts mean to the actors.

Box 3.5

The scientific study of religion

According to Berger (1974: 125):

> The scientific study of religion must bracket the ultimate truth claims implied by its subject. . . . If science means anything at all, as distinguished from other types of mental activity, it means the application of logical canons of verification to empirically available phenomena. And whatever else they may be or not be, the gods are not empirically available, and neither their nature nor their existence can be verified through the very limited procedures given to the scientist. What is available to him is a complex of human experience and thought that purports to refer to the gods.

---

denies the possibility of the existence of non-empirical facets of religion, is inconsistent with the scientific study of religion (although this approach has not become widespread within sociology more broadly).

Nonetheless, both the phenomenological method and the disciplinary distinctiveness of religious studies have been subject to critique. It is argued that the 'methodological agnosticism' (agnosticism about the truth claims or origins of religions) is, after all, a theological standpoint, since it treats religion as *sui generis* (a thing of its own type, which cannot be explained in terms of other phenomena), implying an acceptance of the claims made by believers (Fitzgerald 2000; McCutcheon 1997, 1999, 2001). Thus, critics argue that it is not clear why religious studies should therefore remain a separate discipline. Moreover, religious studies scholars have been challenged for simply reproducing the accounts of 'insiders' as just in need of description and scholarly systematization, rather than exhibiting awareness that religious views may be powerful social, political and ideological tools that are often contested (Oberoi 1994; McCutcheon 1997: 4; Shaw 1995: 70). Shaw, for instance, praises religious studies for developing an interpretive or hermeneutical approach when other disciplines were still 'working through their positivist phases' (1995: 67), but reminds us that 'whose subjective experience is being empathized with is crucial' (ibid.: 67). All too often this is a 'collective subject' who is typically male. Thus, a gender analysis of religion, for instance, is undermined by a phenomenological method if it only aims to develop an understanding of religion from the point of view of the 'collective subject'. As Shaw writes, 'the history/phenomenology of religions is an apt illustration that a hermeneutic of empathy and experience is far from being automatically feminist' (ibid.: 67).

The phenomenological method has, however, been important in carving out a disciplinary distinctiveness for religious studies (Sutcliffe 2004: xxii) in terms of the setting up of academic departments, professional associations and dedicated journals, although today its influence is probably less strong in terms of providing an identity for the discipline. The theory and practice of religious studies today is less likely to view religious traditions as pre-existing and ahistorical phenomena that can be 'discovered' by the researcher and then described to other interested parties, than is a view of religion as a '"social formation". . . embedded in and generated by particular cultural and political contexts and essences' (ibid.: xxvii). Religious studies (as well as theology) has played a vital role in keeping academic interest in the study of religions alive and continues to play a role in empirical and theoretical research about religion that has become more salient in recent years. By contrast, in all of the social sciences, with the exception of anthropology, research on religion has typically taken a back seat, although this neglect is slowly beginning to change. This marginalization of the study of religions can be explained in terms of an assumption that religious traditions both are anti-modern and will disappear, or at least become a matter for private reflection only, as societies develop and modernize (see the discussion of 'secularization' later in the chapter).

This discussion indicates that how we choose to define and conceptualize religion is not a neutral exercise, but is skewed towards particular ways of seeing the world and gives rise to questions about legitimacy, authority and power relations. In addition, both the discussion about definitions of religion at the start of this chapter and the above examination of disciplinary approaches to the study of religion cast doubt upon the suitability of the familiar conceptual tools that we bring to the study of religion, including the very notion of the category 'religion' itself. In the following section of this chapter we will examine this critique in more detail, particularly exploring challenges to the so-called 'world religions paradigm'.

The construction of religion as a category and the emergence of the 'world religions paradigm'

The dominance of the 'world religions paradigm'

What most of the above disciplinary approaches to the study of religion have in common – despite disagreeing about how to study religion, what sort of thing it is and how to define it – is the premise that there is a thing called 'religion'. In fact, this seems so self-evident that it appears to be counter-intuitive to question it. Yet since the 1960s, increasing numbers of scholars have done just that, arguing that the category of religion is a western Christian-influenced 'invention' that emerged following the Protestant Reformation, the rise of science and the Enlightenment in Europe, and that with colonialism and modern globalizing processes it is now taken for granted as having an independent and universal existence. This does not mean to say that the things we label as religion are illusory or that people do not believe in gods and other forms of the super-natural, but that the way we think about these things and their relationship to the so-called secular realm is deeply influenced by a particular view of what a religion should be and how it operates.

More specifically, the very idea of neatly classified 'world religions' (e.g.

Hinduism, Buddhism, Christianity) that have a linear individual history and that can be sharply differentiated from other 'world religions' by virtue of the nature of their distinct beliefs and practices is considered by many contemporary scholars of religion to be a relatively recent (and originally western) invention (Smith 1991 [1962]; Fitzgerald 2000; Asad 1993, 2003; Beyer 2006). Within the 'world religions paradigm', 'world religions' are viewed as originating in particular locations and then spreading to different parts of the globe (normally via migration and conversion): this affords them their 'world' status. This is a paradigm that is widely shared by both scholars and religious practitioners, and has come to shape the way in which people across the globe view religious traditions. In contrast to 'world religions', traditions which do not typically seek converts and which are more likely to be found in a particular location are sometimes called 'ethnic religions'. Both Hinduism and Judaism have been viewed as 'ethnic religions', being centred on particular locations (e.g. India and Israel respectively) and generally not seeking converts (one is born into an 'ethnic religion'). However, these classifications are not clear-cut, with both Hinduism and Judaism now widely found across the globe (although normally more as a result of migration than of conversion) and one is just as likely to find them classified as 'world religions'. Another popular classification is that of 'indigenous religions' (such as African traditional religion or forms of so-called paganism), which continue to be practised in many different localities and often pre-date the 'world religions'. Where these have spread to different parts of the globe, this is more likely to have been via migration than conversion. Often implicit within the 'world religions paradigm' is that the 'world religions' are superior to 'indigenous religions', which are often viewed as primitive, superstitious and polytheistic.

The 'world religions paradigm' has been criticized for distorting features of the traditions we wish to study and understand, since it prioritizes a particular model of religion. As is illustrated in Box 3.6, the extent to which this paradigm is applicable to Hinduism, for instance, has been much debated and it is suggested that Hinduism is being 'shoehorned' into the 'world religions' model. As Oberoi (1994) has demonstrated in his work, prior to the colonial period the distinction between Hinduism and Sikhism, for instance, was far less clear, with distinct religious boundaries only emerging in response to the colonial presence (Plates 3.1 and 3.2). One problem with the dominance of the 'world religions paradigm' (with respect to Hinduism in India, but also more widely) is that, as Geaves writes, it 'does not do justice to the full complexity of religious traditions in the [Indian] subcontinent, as it ignores the eclectic or syncretistic nature of religious experience in that part of the world' (1998: 2). According to the 'world religions paradigm', 'world religions' demand exclusivity, and while conversion is normally a possibility, people are not permitted to 'belong' to more than one religion at a time. However, the existence of styles of 'religious syncretism' in many parts of the globe, where the boundaries between different religious traditions are less clear-cut, points to limitations within the 'world religions paradigm'.

Box 3.6

Is Hinduism a 'world religion'?

'The very name Hinduism owes its origin to chance; foreigners in the West extending the name of the province of Sindh to the whole country lying across the Indus River and simply calling all its inhabitants *Hindus* and their religion *Hinduism*' (Klostermaier 1994: 30–31).

'The term "Hinduism" is, in my judgment, a particularly false conceptualization, one that is conspicuously incompatible with any adequate understanding of the religious outlook of Hindus . . . the classical Hindus were inhibited by no lack of sophistication or self-consciousness. They thought about what we call religious questions profusely and with critical analysis. But they could not think of Hinduism because that is the name we give as a totality to whatever it might be that they thought, or did, or thought worth doing' (Smith 1991 [1962]: 61).

'Is there really a single ancient religion designated by the catch-all term "Hinduism" or is the term merely a fairly recent social construction of Western origin?' (King 1999: 146).

Scholars have argued that prior to the late eighteenth century there was no such thing as a 'world religion' called 'Hinduism'. This is not to imply that a new religion as such was formed in the eighteenth century, but instead to draw attention to the ways in which the colonial 'orientalist' scholars, missionaries and administrators gave a name (Hinduism) to the totality of diverse practices found in India followed by the majority of the population (and that were not perceived to be associated with other religions). Similar processes can also be demonstrated for Sikhism and Buddhism, for instance (Oberoi 1994; Almond 1988). In naming Hinduism, however, the beliefs and practices associated with the priestly Brahmin caste (e.g. the elite Sanskrit tradition) were focused upon rather than the colloquial village or popular 'religion' that most people actually engaged in. Popular religion was generally regarded as an aberration, a corruption of an authentic religious tradition that reflected a golden age of human civilization (e.g. the religion of the Aryan authors of the Vedas, the most ancient 'Hindu' texts). Indigenous thinkers and movements also played a part in this process (e.g. the Brahmo Samaj formed in 1828 by Ram Mohan Roy). As Killingley notes, Ram Mohan Roy was (probably) the first Hindu to use the term 'Hinduism', in the early nineteenth century (1993: 62–63). This construction of religious boundaries, however, is not merely a colonial interest; it has also been central to the emergence of distinct religio-nationalistic identities in India (e.g. Sikh and Hindu nationalist movements – see Box 4.8 on p. 101).

Source: adapted from Tomalin (2009b).

The challenge of 'religious syncretism' to the 'world religious paradigm'

Religious syncretism can refer to slightly different yet related scenarios, including the practice of more than one religion simultaneously; the incorporation of some aspects of one religious tradition into another; or the blending of different religious traditions to form a new one. Syncretism is not found only

Plate 3.1 *A public prayer in the River Ganges in the Indian holy city of Haridwar*

Source: http://www.flickr.com/photos/lazyoldsun/140265364/Creative Commons.

Plate 3.2 *Sikh pilgrim at the Harmandir Sahib (Golden Temple) in Amritsar, India*

Source: Paulrudd (Creative Commons). Online, available at: http://wikimediafoundation.org/wiki/File:Sikh_pilgrim_at_the_Golden_Temple_(Harmandir_Sahib)_in_Amritsar,_India.jpg.

The man pictured has just had a ritual bath.

in non-western religions or contexts: Christianity in Europe, for instance, has 'incorporated' many pagan practices and festivals. However, as Maroney writes:

> Syncretism has often been the sworn enemy of the three great Abrahamic monotheistic religions of the West [Islam, Christianity and Judaism]. Whereas other religious traditions often openly adopt the practices, customs, styles of worship and even the deities of other religions, the Abrahamic faiths have infrequently acknowledged the debt they owe their predecessors and contemporaries.
>
> (2006: 6)

Traditionally, members of the Abrahamic religions consider that these are 'revealed faiths', which means that the beliefs, texts and practices of their religion are held to emanate directly from God. In these cases the term 'syncretism' is viewed as negative or disparaging for suggesting that the religion has deviated from the 'true path' by absorbing heretical elements. Thus, what may sometimes be a useful analytical term to account for the shape and formation of different religious traditions can be contentious in other settings. Moreover, there are cases where the purging of perceived syncretic influences in religious traditions results in violence and discrimination. The spate of violence at and destruction of Sufi shrines in South Asia and Africa (Sufism is a branch of Islam that is marked by its syncretic or hybrid nature) by Islamists (who promote a 'pure' Islam) during 2012 is one such example.

Despite the negative connotations of the term, particularly within Abrahamic religions, much religious activity in developing countries is marked by its syncretic nature. Thus, if we are seeking to locate statistics that indicate the proportion of a population that practice one religion or another, these may well be misleading since they suggest that people only 'belong' to one tradition (Pew Forum 2010). Critics of the 'world religions paradigm' argue that distinct boundaries between religious traditions are a feature of the western and Christian-influenced model of religion. Although this model has now become superimposed upon 'religions' in the non-West, these have not made the transition to 'world religions' completely. For instance, we find syncretism within religions in India, including Hinduism, Islam, Sikhism and Christianity, which not only overlap with each other but also often include elements of tribal (*adivasi*) religion. As Box 3.7 illustrates, this is also the case in Africa, where people often practise African traditional religions (ATR) alongside Christianity or Islam. It is interesting to note, however, that many contemporary Africans tend to see ATR as related to cultural practices (although ATR has many of the characteristics of religion) and to reserve the word 'religion' for Islam and Christianity, which perhaps reflects a desire to avoid the appearance of engaging in syncretic or hybrid forms of religiosity.

In the next section we will look more closely at critiques of the 'world religions paradigm' that aim to explain its origins as well as to demonstrate the

Box 3.7

Islam and Christianity in sub-Saharan Africa

According to the Pew Forum (2010: 1):

> The vast majority of people in many sub-Saharan African nations are deeply committed to the practices and major tenets of one or the other of the world's two largest religions, Christianity and Islam. Large majorities say they belong to one of these faiths, and, in sharp contrast with Europe and the United States, very few people are religiously unaffiliated. Despite the dominance of Christianity and Islam, traditional African religious beliefs and practices have not disappeared. Rather, they coexist with Islam and Christianity. Whether or not this entails some theological tension, it is a reality in people's lives: Large numbers of Africans actively participate in Christianity or Islam yet also believe in witchcraft, evil spirits, sacrifices to ancestors, traditional religious healers, reincarnation and other elements of traditional African religions.

consequences it has had for how we view different styles of 'religion' across the globe.

Interrogating the 'world religions paradigm'

As early as 1962, Wilfred Cantwell Smith in his famous book *The Meaning and End of Religion* wrote that scholars should stop using the terms 'religion' and 'religious', 'since the *reification* they imply makes them "confusing, unnecessary, and distorting"' (Beyer 2006: 62, citing Smith 1991 [1962]: 50). Jonathan Z. Smith, a religious studies scholar writing twenty years later, expresses similar sentiments, 'claiming that "religion is solely the creation of the scholar's study" and "has no independent existence apart from the academy"' (Beyer 2006: 62, citing Smith 1982: xi). Reflecting the influence of postcolonial studies in particular, this argument now also carries with it a greater awareness of the construction of the category 'religion' as having ideological and political ramifications (e.g. Asad 1993; Beyer 2006, Fitzgerald 1990; McCutcheon 1997; Chidester 1996).

I shall next briefly outline approaches to this debate from different disciplinary perspectives: postcolonial anthropology (Talal Asad), international relations (Scott Thomas) and sociology (Peter Beyer). While each scholar's account emphasizes his particular disciplinary concerns, each arrives at the similar conclusion that the concept of religion, including the idea of discrete 'world religions', is modern (i.e. post-Reformation), western and Christian influenced. Each adopts a position that is radically opposed to earlier positivist tendencies within the social sciences, which assume that there are 'real' social phenomena that can be observed by neutral researchers and the laws governing them identified and described. Instead, they approach their inquiry from a position that contextualizes, historicizes and interprets the meaning of the taken-for-granted category 'religion' from a postmodern or postcolonial perspective.

A perspective from postcolonial anthropology

Talal Asad, an anthropologist working from within the postcolonial tradition and influenced by the work of the French theorist Michel Foucault, is one such critic of the category of religion. He adopts a genealogical method, following Nietzsche and Foucault, which traces the origins of taken-for-granted words and concepts in order to deconstruct their essential nature. Asad argues that 'there cannot be a universal definition of religion, not only because its constituent elements and relationships are historically specific, but because that definition is itself the historical product of discursive processes' (1993: 29). For Asad, there can be no *transhistorical* definition of religion because different definitions have existed at different times (and indeed in different places). The fact that we now tend to act as though religion has an essence that can be identified and defined across cultures is, he argues, the product of western ways of viewing religion that had emerged by the seventeenth century, following the Protestant Reformation in the sixteenth century and the devastating wars of religion in the late sixteenth and early seventeenth centuries (ibid.: 40). He writes that at this point we find

> the earliest systematic attempts at producing a universal definition of religion ... in terms of beliefs (about a supreme power), practices (its ordered worship), and ethics (a code of conduct based on rewards and punishments after this life) – said to exist in all societies.
>
> (ibid.: 40)

In an era of colonial expansion and missionary activity, being able to identify religions in new locations as they were encountered became crucial: knowing what a religion was became necessary. Moreover, there was an emphasis on belief, which meant that 'religion could be conceived as a set of propositions to which believers gave assent, and which could therefore be judged and compared as between different religions and as against natural science' (Asad 1993: 41). Thus, as Asad argues, 'from being a concrete set of practical rules attached to specific processes of knowledge and power, religion has come to be abstracted and universalized' (ibid.: 41, 42). Moreover, this process of naming and defining religions, alongside an emphasis upon belief, signals the emergence not only of the modern category 'religion' but also of the 'world religions paradigm', which promotes the view that we can identify distinct religious traditions that can be defined as part of a general category 'religion' but differ from one another in terms of their beliefs.

In a later book, Asad (2003) focuses on the ways in which the split between the 'religious' and the 'secular' is the product of particular historical processes. These include the Protestant Reformation, which removed the all-encompassing power of the Catholic Church (as a reaction against its wealth, corruption and support for feudalism); the rise of science (which challenged the role of the Church in providing an overarching source of meaning and explanation, including the rise of reason and empirical observation above blind faith and revelation); and the rise of individualism and the nation-state. Box 3.8 provides a definition of the term 'secular', which suggests that the meaning and use of the word developed over time, eventually becoming applied to the split between the religious and the secular realms. Thus, the impact of the historical processes outlined above was to effect the gradual separation of these two realms – the religious and the secular – such that they each had their own sphere of operation. According to Asad, however, 'although religion is regarded as alien to the secular, the latter is also seen to have generated religion' as we understand it today (2003: 193).

Asad also extends his account to look at the meaning and emergence of 'secularism' (see Box 3.8 for two related definitions of secularism). He undertakes an 'anthropology of secularism' with the particular aim of better understanding the ways in which Muslims are treated within western secular states. The preference for 'secularism' (here understood as the political ideology that religion should be kept separate from other domains of life) has come to dominate western, particularly European, politics. One of the functions of the secular nation-state is to protect society from the perceived irrationality of

---

Box 3.8

Secular and secularism

The term 'secular' is derived from the same etymological root (L. *saeculum*) as the French word *siècle*, meaning 'century', or 'age'. . . this derivation is evident in the astronomical use of the word secular to talk about processes of change over long periods of time, such as changes in planetary orbits. However, most people are perhaps more familiar with the usage that originated in early ecclesiastical texts, in which the term 'secular' was commonly opposed to 'regular'. 'Regular', in this context, was a term applied to those persons subject to the role of a monastic order. Thus, its opposite – 'secular' – was used to denote worldly affairs.

(Fox 2005: 292)

According to *Encarta*, 'secularism' comprises a commitment to bringing about or maintaining the 'secular':

1. exclusion of religion from public affairs: the belief that religion and religious bodies should have no part in political or civic affairs or in running public institutions, especially schools
2. rejection of religion: the rejection of religion or its exclusion from a philosophical or moral system.[1]

religion, as well as from any conflicts caused by religiously informed politics and the potential inequalities generated via socially influential religion. Thus, religion becomes partitioned off, differentiated from other spheres of life and relegated to the private lives of individuals and the day-to-day functioning of religious establishments. The effect has been to define religion in a particular way, as that which exists in the private sphere, where it can be disciplined and controlled by the secular nation-state. Hence, manifestations of religion that do not fit this model are perceived as problematic and threatening. In deconstructing the taken-for-granted distinction between the secular and the religious, Asad aims to account for the ways in which the West thinks about Islam, tending to treat Muslims as outsiders and 'as a distinctive minority to be tolerated (the liberal orientation) or restricted (the national orientation), depending on the politics of the day' (Connolly 2006: 76). Asad argues that while the modern nation-state holds the view that the secular sphere is a neutral domain of non-coercion, truth and equality, in reality it carries with it certain prejudices about different (non-Christian) religions as well as strong views about what a religion should be.

A perspective from international relations

Scott Thomas's analysis shares many similarities with that of Asad, since he is also interested in interrogating the ways in which modern western liberal societies view the mixing of religion and politics – part of the so-called 'resurgence of religion' – as inevitably 'violent and dangerous to reason, freedom, and political stability' (2005: 21), therefore requiring the neutral secular state to step in. Thomas argues that this view relies upon a modern concept of religion (and the secular) that 'was invented as part of the political mythology of liberalism and now has emerged as a universal concept applicable to other cultures and civilizations' (ibid.: 21). Prior to the rise of the secular state and nationalism, he argues, religion was understood as a 'community of believers' rather than the modern meaning of a 'body of beliefs'. Over time, the meaning of religion changed as it became focused on privately held beliefs, and the Catholic Church, once the overarching source of social coherence and legitimacy, lost its monopoly to the secular state. This view of religion as concerned with privately held beliefs suggests that the proper place for religion is in the private sphere of society rather than mixing with politics and other matters of the state in the public domain. This 'modern' understanding of religion and the role of the secular state, Thomas writes, has consequently distorted our reading of the role of religion in wars of the past, as well as 'in ethnic and internal wars or in international conflicts today' (ibid.: 23).

The wars of religion in early modern Europe (the civil war between Protestants and Catholics in France (1550–1650) and the Thirty Years War (1618–1648)) are commonly viewed as examples of what happens when religion is politicized and attempts to defend or impose its doctrines and beliefs against those of other religions, therefore seeming to reinforce the need for the secular state to step in. Thomas argues, however, that this presumes a particular modern concept of religion as focused upon privately held beliefs, which although emergent did not yet exist (see also Cavanaugh 1995). Therefore, the so-called wars of religion were not actually fought over religion, as we understand it today, suggesting perhaps that it is anachronistic to call them wars of 'religion' (Cavanaugh 1995; Holt 1993). Instead, these wars should be viewed as concerned with preserving religion as a 'community of believers' and not with clashing religious doctrines (Thomas 2005: 23–26). Thomas suggests that each community was attempting to 'define, redefine, or defend' its 'sacred notion of the community' and 'the social boundaries between the sacred and the profane as a whole' (ibid.: 24). Eventually, the Treaties of Westphalia (1648) brought the Thirty Years War to an end, and a new formulation between the state and religion emerged in which, as Thomas explains, 'the concept of Christendom itself was discredited and . . . [t]he new Westphalia state system recognized the state as the dominant actor, replacing the transnational authority of the Catholic Church' (ibid.: 54).

What is the significance of these observations about the 'myth' of the wars of religion for contemporary approaches to religion in international relations? First, Thomas does not deny that religious communities have engaged and continue to engage in acts of war but is concerned that the reasons for those wars are often misunderstood, and that this misunderstanding has led to a continuing misinterpretation of the causes of contemporary religious wars. Understanding why conflict occurs is a central theme of international relations and is necessary in order to be able to suggest solutions and to help prevent similar wars in the future. Second, the understanding of religion as 'a body of beliefs', privately held and separate from the secular state, has come to shape the way in which western liberal societies interpret religion and has implications for understanding the global 'religious resurgence': religion is typically viewed as a problem and something to be monitored and disciplined.

For Thomas, 'the problem with applying the modern concept of religion to the study of many of the societies in central Europe, central Asia, and most of the non-Western world is that they have still not entirely made, or are struggling not to make, this transition to a modern concept of religion' (2005: 26). While this statement might seem to embody a colonialist value judgement – that we think they *should* make this transition to accommodate the 'world religions paradigm' – it could also be taken to imply that this transition is inevitable. In many societies across the globe, religion is not marginalized or privatized and continues to define community boundaries in the public sphere. Instead of being about privately held religious beliefs, religion is as much about identity and community, which frequently (even in the secular West) means that it appears to spill into the so-called public sphere. This is not a failing of the secular state per se, or indeed something to be necessarily worried about, but instead suggests that religion is possibly not what we have typically understood it to be. International relations, and indeed other academic disciplines, need to take religion seriously and also re-vision the way that they think about and approach the study of religion. This change of outlook not only presents a challenge to the political mythology of liberal modernity but also necessitates the creation and adoption of new or modified frameworks for thinking about the place of religion in modern global society. As Thomas writes:

> The global resurgence of religion . . . challenges the idea that secular reason can provide a neutral stance from which to interpret religion, and it opens up the possibility of multiple ways of being 'modern,' making 'progress,' or being 'developed' consistent with a variety of cultural and religious traditions.
>
> (ibid.: 21)

A perspective from sociology

Finally, we will examine the interrogation of the 'world religions paradigm' offered by Peter Beyer, a sociologist who has been one of the main commentators on the relationships between religions and globalization (1994, 2006). Like Asad and Thomas, he highlights the importance of shifts taking place in European society around the time of the Reformation. Employing a systems theory approach that draws on the work of the social theorist Niklas Luhmann, he considers that the shift from medieval to modern societies has been marked by the radical differentiation of different functional 'social subsystems' from one another: for instance, science, politics, medicine and religion. This process of differentiation has intensified over time and he views modern global society as constituted by radically differentiated subsystems that aim to have dominance over their particular spheres of action. While the differentiation of 'religion' as a realm separate from the other social subsystems gives it a distinct profile, it also raises questions about what its proper sphere of action should be and gives rise to attempts by 'non-religion' to contain it, as well as internal disputes within religions about what is orthodox and what is not. Thus, 'religion' becomes differentiated not only from other social subsystems but also from other religions, and while these processes of religious differentiation are global, and have led to the emergence of the so-called world religions, the exact form that they take in particular locations differs. For instance, Christianity and Islam have adapted more easily to differentiation, both for historical reasons and because of the nature of those traditions, whereas Hinduism 'emerged' much later and, as Thomas also notes, has not quite made the transition to the modern understanding of religion (2005: 26). According to Beyer, the pattern is different again in East Asia, where there has been and continues to be much resistance to the seemingly 'religious' being viewed as religion, for instance with respect to Confucianism in China and Shinto in Japan. He writes:

> [R]eligion is not just a social institution that exists more or less in all societies and all periods of human history, but only takes on different forms in different circumstances. Form and content are not so easily separable. Instead it is the modern form that religion has taken, its social differentiation precisely as religion, that makes it seem almost self-evident to us as observers that it has always and everywhere been there *in some form*.
>
> (2006: 299; emphasis in the original)

He argues that the various ways that religions have adapted to differentiation (both from other social systems and from one another) shapes how they operate and 'naively applying the concepts for understanding religions in the West to these regions is both risky and more or less inappropriate' (ibid.: 251).

Limitations of familiar concepts in the study of religion

As the preceding discussion has demonstrated, the familiar concepts that we bring to the study of religion may be limited in their ability to help us understand religious phenomena globally in a number of ways. First, if the conventional definitions of religion, as well as the very category itself, and the tendency to think in terms of discrete 'world religions', are contingent to a particular modern and western context, then these tools may be limited in their ability to throw light on 'religious' phenomena in other contexts. As Fitzgerald writes:

> I am particularly concerned with the use of the word 'religion' in studies of non-western value systems and cultures. I believe that it is in these contexts that we can most easily see how a western concept . . . is continually being foisted on non-western societies even though its application is so obviously problematic.
>
> (2000: 9)

Nonetheless, this important observation is complicated by the fact that the term 'religion' and the idea of discrete 'world religions' are in wide use globally and, even if they are a fairly recent invention, they have come to influence the ways in which all people across the globe think about their 'religions'. In a pragmatic sense the possibility of abandoning the idea of 'world religions', or even the term 'religion', does not seem feasible. Arguably, what these debates suggest is not so much that there is no 'religion' or 'world religions' to be studied and researched, but that the way these religions are defined and understood varies between different contexts and as researchers we cannot assume that the religions that are the focus of our study will resemble the western secular Christian model. We need to be *suspicious* about the 'world religions paradigm' when it suggests that all religious formations are rather more like the western Christian model than they actually are.

Second, the 'world religions paradigm' not only presumes a distinction between different religions but also relies upon the distinction between the religious and the secular, and this is not always easy to find in many developing contexts. Sometimes we hear the phrase that a certain religious system is not a religion after all, but instead a 'way of life'. The import of this statement is to emphasize the extent to which it is not always possible to say what is religious and what is not, where religion strongly impacts upon almost every aspect of life. Box 3.9 illustrates this with respect to Hinduism and draws attention also to an understanding of the Hindu tradition as not being like 'other religions in the world' since it does not seem to fit the 'world religions paradigm'.

Box 3.9

Hinduism as a 'way of life'

In 1995, Chief Justice P. B. Gajendragadkar was quoted as follows in an Indian Supreme Court ruling:

> Unlike other religions in the world, the Hindu religion does not claim any one prophet; it does not worship any one god; it does not subscribe to any one dogma; it does not believe in any one philosophic concept; it does not follow any one set of religious rites or performances; in fact, it does not appear to satisfy the narrow traditional features of any religion or creed. It may broadly be described as a way of life and nothing more.

Source: www.brahmakumaris.info/w/index.php?title=Hindu (last accessed 13 September 2012).

Third, the terms 'secular' and 'secularism' can have different meanings in different contexts. In India, for instance, instead of 'secular' denoting 'worldly affairs' as opposed to 'religious' ones (and by implication a rejection of religion), when applied to politics the term 'secular' refers to a situation where the state does not show preference for one religion over another (Brass 2006). In a country where religion is seen as a 'way of life' (this is arguably as true in important ways for other religions in the country: Islam, Christianity, Sikhism, Buddhism and Jainism), the possibility of a secular state in the western sense is meaningless. Thus, the European model of secularism has mutated as it has spread to different locations.

Neither, however, are secular states in the Global North secular in a single 'western' sense. There are a variety of secularist formations between Europe and North America (see, for instance, de Beaufort *et al.* 2008). In the United Kingdom, for example, while the Church of England is the state religion, headed by the British monarch, the link between church and state is largely nominal and the country is governed in a secular fashion. In France, by contrast, there is a strict separation between church and state (known as *laïcité*), which has involved the banning of religious apparel and displays in schools in 2004 and more recently the ban on face veils in 2011. In the United States, which also has a constitutional separation between church and state, this has not taken what many consider to be the more extreme route in France. Instead, we often find politicians in the United States, particularly those from the more right-wing Republican Party, expressing religious faith alongside their political commitment.

The existence of variations in secularist formations in the Global North is also suggested by the fact that there is no straightforward correlation between a country's being constitutionally 'secular' and actual levels of religiosity among the population. For instance, the citizens of the United States are highly religious compared to those in Sweden, even though both are 'secular' states. While Sweden ranks as one of the most secular countries in Europe, with only a

Table 3.1 *Percentage of people who consider that religion is 'very important'*

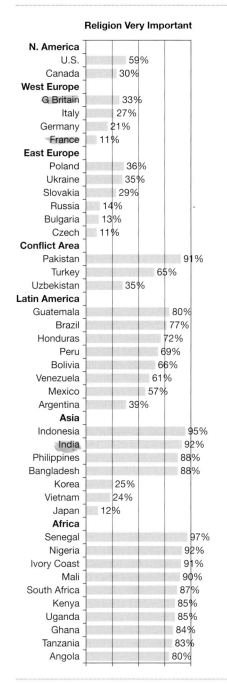

Religion Very Important

Region / Country	%
N. America	
U.S.	59%
Canada	30%
West Europe	
G Britain	33%
Italy	27%
Germany	21%
France	11%
East Europe	
Poland	36%
Ukraine	35%
Slovakia	29%
Russia	14%
Bulgaria	13%
Czech	11%
Conflict Area	
Pakistan	91%
Turkey	65%
Uzbekistan	35%
Latin America	
Guatemala	80%
Brazil	77%
Honduras	72%
Peru	69%
Bolivia	66%
Venezuela	61%
Mexico	57%
Argentina	39%
Asia	
Indonesia	95%
India	92%
Philippines	88%
Bangladesh	88%
Korea	25%
Vietnam	24%
Japan	12%
Africa	
Senegal	97%
Nigeria	92%
Ivory Coast	91%
Mali	90%
South Africa	87%
Kenya	85%
Uganda	85%
Ghana	84%
Tanzania	83%
Angola	80%

Source: taken from a 44-nation survey carried out under the Pew Global Attitudes Project: 'Americans' views are closer to [those of] people in developing nations than to the publics of developed nations' (Pew Research Center 2002).

small percentage regularly participating in religious activities, 39 per cent of Americans state that they attend 'services' at least weekly, and 58 per cent say that they 'pray' daily (Pew Forum 2008). The United States is often singled out for contradicting a straightforward relationship between modernization and secularization, since it has relatively high numbers of people who are religious despite its being a wealthy (and indeed 'secular') nation. As Table 3.1 illustrates, according to the Pew Global Attitudes Project, poorer nations tend to be more religious, with the United States being the exception. Thus, despite the assumption that modernization and secularism in the West would lead to increased secularization, this has not occurred in the way predicted.

This brings us to our final limitation of the familiar concepts that we bring to the study of religion: challenges to the universality and normativity of secularization theories. The gradual distancing of the religious from the secular since the sixteenth century, and the rise of secularism as the Enlightenment response to dealing with the dangers implicit in religious engagement with public or political affairs, carried with it an assumption that religion would increasingly retreat to its own sphere of action as a matter of conscience in the private lives of individuals. Indeed, it was thought also that it might well disappear altogether as people acquired alternative 'modern' ways of viewing the world and responding to its challenges. This focus on secularization – as an inevitable part of the modernizing development process – has been critiqued not only for reflecting a western normative position but also for going against the evidence. Rather than withering away, religion seems to be growing in strength in some parts of the globe, both developed and developing, and theories of secularization now appear to be simplistic and limited in their ability to account for the perseverance or even resurgence of religion. Instead, I shall suggest that it is more fruitful to refocus our attention to thinking about *religious change*, since processes of secularization are only part of the story.

Problematizing secularization theories

As Gorski writes, it was during the 1960s that the '"secularization thesis" was integrated into "modernization theory" and became one of its central axioms. As societies modernized, they became more complex, more rationalized, more individualistic – and less religious' (2003: 111). However, the roots of secularization theory go back further, to

> the early nineteenth century and the writings of Henri Saint-Simon and Auguste Comte . . . [who] argued that human history passes through a series of distinct stages, in which the power and plausibility of traditional religion are gradually and irreversibly undermined by the growing influence of the state and of science.
>
> (ibid.: 111)

Thus, classical European sociology in the nineteenth and early twentieth centuries not only predicted the decline of religion but also perceived it as being at odds with modernity. The classical social theorists lived in a world where there was already a great deal of differentiation between the secular and the religious, and they considered that the role of religion in providing 'social cohesion' (Durkheim) or 'meaning' (Weber) was rapidly diminishing. In the post-Second World War period, sociology became more widely practised and studied in the West and the secularization thesis became popularized in the writings of sociologists such as Peter Berger, Thomas Luckmann, Harvey Cox and Bryan Wilson in the late 1960s and 1970s.

By the 1980s, however, the dominance of secularization theories had begun to recede and today they are largely viewed, if not as wrong, as seriously flawed and oversimplified. This is particularly the case for positions 1, 2 and 3 in Box 3.10, which outlines different positions on secularization within the sociological literature. In particular, these cannot account for patterns of religiosity in much of the non-West, as well as high levels of religiosity in the United States; they cannot accommodate the 'religious resurgence' globally; and their normative quality (i.e. that modernization and religious decline would or should go hand

---

Box 3.10

Secularization

Woodhead and Heelas (2000) identify four main positions on secularization within the sociological literature:

(1) *The disappearance thesis*: religion is in decline and will eventually disappear.

> The modern West has produced an increasing number of individuals who look upon the world and their own lives without the benefit of religious interpretations . . . we have enough data to indicate the massive presence of . . . [the secularization of consciousness] in the contemporary west.
>
> (Berger 1967: 108)

(2) *The differentiation thesis*: religion has become relegated to the differentiated private sphere and has no place in the secular, public sphere.

(3) *The de-intensification thesis*: religion will become less important over time but will not disappear completely. Secularization does not

> predict the disappearance of religiosity, not even of organized religion; it merely indicates the decline in the significance of religion in the operation of the social system, its diminished significance in social consciousness and its reduced command over the resources (time, energy, skill, intellect, imagination, and accumulated wealth) of mankind.
>
> (Wilson 1985: 14)

(4) *The coexistence thesis*: religion and non-religion will exist side by side.

in hand) is seen by many as being the product of the imposition of a western perspective.

What the failure of the secularization thesis suggests is not that secularization never happens but that it is only part of the story. Instead, what we actually witness are diverse and widespread processes of religious change that feed from and influence broader processes of social change. In some places, religion becomes less strong and in other places it increases in significance. In fact, over the past couple of decades secularization has received a far more sophisticated treatment by scholars seeking to understand, among other things, the rise of public religion in the 1980s, specifically because it seemed to contradict the predictions of theories of secularization.

The work of the sociologist of religion José Casanova has been one of the strongest influences in this endeavour. He postulated three different understandings of the term 'secularization': (1) the decline of religious beliefs and practices; (2) the privatization of religion; and (3) the differentiation of the secular spheres, whereby they become free from religious influence. His aim was to interrogate each connotation of the term and to identify the extent to which they were borne out by the evidence. He accepts the third sense, but has been criticized for adopting a western-centric view and in later publications has emphasized that he did not consider secularization in this sense as 'a general universal process of human and societal development' (2008: 102), instead basing his account on the West – specifically Europe, where this can be demonstrated to have occurred. While Casanova also accepts the first sense of secularization – that, overall, religious belief and practice are in decline – it is the privatization of religion that he challenges. He argues that we are witnessing a 'deprivatization' of religion where '*religious traditions throughout the world are refusing to accept the marginal and privatized role which theories of modernity as well as theories of secularization had reserved for them*' (1994: 5; emphasis added). It is this aspect of the secularization thesis which Casanova suggests is most interesting and also the one that is most challenged by the evidence.

While it is not entirely clear that these three understandings of secularization can be so easily differentiated (Asad 2003: 182, for instance, argues that if the 'deprivatization' of religion has occurred, then the other two meanings of secularization must also be invalid), Casanova does systematically explore evidence that the place of religion in modern societies does not seem to be the one predicted for it by theories of modernity. This is also Asad's intention.

Thus, while theories of secularization may be descriptively useful in pinpointing patterns of religious change in certain locations, their normative dimensions are widely discredited. The idea that a particular style of modernization is inevitable as societies 'develop' and that a marker of this will be the decline of religion as a public force is no longer taken for granted in the way that it once was. Moreover, the claim that 'religious resurgence' or the 'deprivatization of religion' somehow signals that *modernity has failed* presumes a particular understanding of modern societies as those where religion is absent from the public sphere. Instead, as Thomas writes, 'what if the global resurgence of religion can no longer be interpreted within the traditional categories of social theory?' (2005: 44). He suggests instead that we need to adopt a 'postmodern' frame of reference:

> What a *postmodern perspective* opens up is the possibility that there may be other ways of being 'modern', making 'progress', or being 'developed'. Although postmodernity can mean a lot of things, in this sense it suggests that rather than there only being one path to modernity – Westernization, there may be multiple paths, 'multiple modernities', or multiple ways of being appropriate to the different cultural and religious traditions in the developing world.
>
> (2005: 45; emphasis added)

The idea that we are now in a postmodern rather than a modern era began to emerge in the mid-twentieth century in the work of philosophers such as Jean-François Lyotard and Jean Baudrillard. In contrast to the emphasis upon rationality and progress that marked modernity, postmodernity is characterized by rapid change and uncertainty, and a loss of faith in progress and universal narratives. Others argue, however, that we are in a later phase of modernity where the underpinnings of the Enlightenment are still firm: we have not yet made a radical break with modern values and ideas. Giddens, for instance, calls this 'high modernity' (1991: 163). Whether one adopts a postmodern frame or not, what there does seem to be some agreement on is the fact we live in an era where the certainties of traditional versions of modernization theory and the earlier promises of modernity now seem less secure. As we saw in Chapter 2, this also appears to be reflected in evolving approaches to how donor-driven development has been understood and approached since the 1950s, from a focus upon an economic model that stressed growth and modernization, to more diverse approaches that increasingly aim to reflect the worldviews of different communities.

Summing up

The aim of this chapter has been to examine concepts and theories for studying religions and to suggest that these may be limited in their ability to account for religious forms in non-Christian and non-western contexts. We have explored the difficulties involved in defining religion, not only because religions are diverse in their manifestations but also because the definitions of religion are only capable of reflecting what scholars of the study of religion have conventionally understood religion to be, reflecting a western colonial and Christian heritage. Our discussion of different disciplinary approaches to the study of religion, highlighting both 'outsider' and 'insider' approaches, indicated that how we choose to define and conceptualize religion is skewed towards particular ways of seeing the world, including the very notion of the category 'religion' itself. We examined a series of critiques that problematize the category of religion and the dominance of the 'world religions paradigm', and draw attention to the emergence of the distinction between the religious and the secular as the product of a particular period within European history, following the sixteenth-century Protestant Reformation in Northern Europe. This has involved transposing a Christian understanding of the nature of religion onto non-Christian and non-western contexts, thereby distorting our understanding of those 'religions' as well as transforming the way in which they come to understand themselves. Moreover, the chapter has demonstrated that not only do our understandings of what a religion is rely upon a particular western perspective, but so do predictions about its future. We have discussed how theories of secularization, which predict the decline of religious belief and practice and/or their relegation to the private sphere, do not seem to fit the evidence provided by the continued or increased significance of religion in many places across the globe, both publicly and privately.

These observations present a number of challenges for studying the relationships between religions and development and require reflexivity in the application of the familiar tools and concepts that we bring to the study of religion. First, religions in developing contexts do not necessarily fit the traditional western Christian-influenced model of what a religion *should be*. For instance, Buddhism does not teach belief in a divine being and Hinduism has a greater focus on 'orthopraxy' than on 'orthodoxy'. Hence, for most Hindus belief and doctrine are secondary to practice and how one acts and behaves. This presents a challenge to attempts to produce a universal substantive definition of religion. Second, the idea of distinct boundaries between religious traditions is a feature of the western model of religion, whereas in many non-western contexts styles of 'religious syncretism' or 'hybridity' are common, where people seem to practise more than one religion at a time. Third, it is often not possible to clearly separate the religious from the secular in the way that western models of religion demand. In highly religious contexts, people may not think about of they do or what influences them as being 'religious'. Neither may apparently religious organizations be clearly labelled as 'faith-based' (see Chapter 8 for a discussion of this).

The avoidance of religion within the mainstream donor-driven development project, since its inception in the post-Second World War period, was underpinned by a commitment to secularism and a belief in secularization. However, there is now a need to rethink the relationship between religions and development, moving from dominant views that religion hinders development and will eventually disappear to a view in which a consideration of religions is given serious attention. The contemporary 'religious resurgence' may actually highlight a 'crisis' or 'failure' of modernity after all, but this is not a failure that can necessarily be corrected by working harder to keep religion at bay (as the secularist model demands). Instead, it means that we need to develop new ways of thinking about the relationships between religions and society that do not rely on the modern categories of religious–secular and private–public. However, what would this entail for donor-led development policy and practice? If we were to follow Wilber and Jameson (1980) in their argument that development needs to connect to the 'moral base' of society to be successful, then to what extent would we need to undertake a radical re-visioning of what development means and how it should proceed? As already noted, mainstream development actors have been criticized for engaging instrumentally with religion (i.e. only where their predetermined goals and methods are promoted). Most governments and donors in the Global North generally think that religion gets in the way of development. While they are likely to consider that it is acceptable when it motivates people to promote development, it is felt to be problematic when it interferes with the secular project of development according to that 'project's own understanding of what constitutes rationality, progress, social justice, and modern economic development' (Thomas 2005: 220).

In the following chapter we shall look at some of the ways in which religious traditions approach wealth, poverty and economic development. We will examine the claim that mainstream development discourse (the 'secular model') tends to emphasize economic development and material wealth above other factors and will explore the extent to which these religious approaches are significantly different from secular approaches.

Questions for discussion

1 What are the main problems encountered when attempting to define religion?

2 In what ways and for what reasons has the category 'religion', and in particular the idea of so-called world religions, been contested?

3 What are the main problems with theories of secularization?

4 How are the debates aired in this chapter relevant for the study of 'religions and development'?

Recommended reading

Beyer, Peter (2006) *Religions in Global Society*, **London: Routledge.** This book looks at religion in modern global society as itself a dimension of the historical process of globalization, drawing upon examples from Islam, Hinduism, African traditional religions and New Age spirituality.

Casanova, José (1994) *Public Religions in the Modern World*, **Chicago: University of Chicago Press.** This classic text reconsiders the relationship between religion and modernity, problematizing the privatization of religion in particular.

Fitzgerald, Timothy (2000) *The Ideology of Religious Studies*, **New York: Oxford University Press.**

McCutcheon, Russell T. (1997) *Manufacturing Religion: The Discourse on Sui Generis Religion and the Politics of Nostalgia*, **New York: Oxford University Press.** The above two texts challenge the idea of the category of religion and of religious studies as a separate discipline.

Woodhead, Linda and Paul Heelas (eds) (2000) *Religion in Modern Times: An Interpretive Anthology*, **Oxford: Blackwell.** A useful anthology of over 300 important readings on religion in modern societies, with interpretive introductions to each themed section.

Websites

http://hirr.hartsem.edu/ency/ This is the online version of the *Encyclopedia of Religion and Society*, edited by William H. Swatos Jr and published by Altamira Press (1988).

http://pewforum.org The Pew Research Center's Forum on Religion and Public Life was launched in 2001 and seeks to promote a deeper understanding of issues around the intersection of religion and public affairs. It conducts surveys, demographic analyses and other social science research on religion and public life in the United States and globally. The website provides a large number of publications, containing data and analysis.

4 Religious approaches to development

This chapter will:

- address the questions: *How might members of religious traditions approach 'development'? In what ways does religion influence visions of development as well as the policies and strategies, to reach those visions? Are these religious approaches significantly different from secular approaches?;*
- highlight the diversity within religious traditions and the difficulties this poses for answering the above questions;
- examine the claim that mainstream development discourse (the 'secular model') tends to emphasize economic development and material wealth above other factors;
- look at the ways in which members of three of the major 'world religions' (Islam, Hinduism and Christianity) have traditionally viewed wealth and poverty, and how their outlook might shape their approach to contemporary debates about economic development.

Introduction

A recurring discourse within debates about religions and development is the extent to which 'secular' development approaches are deficient because they tend to emphasize economic and material well-being at the expense of other aspects of human life, such as 'spiritual' development, aesthetic pleasure or community cohesion – considerations for the so-called affective dimensions of life, and ones that are generally important within religious traditions. Membership of a religious community, moreover, potentially offers support systems that may be drawn upon in difficult times and can also provide moral frameworks that promote the 'right' way to act (i.e. according to the teachings of that tradition) in day-to-day activities as well as in more challenging situations. Beyond these general qualities of religions, however, many members of religions also consider that their traditions have something distinctive to offer in terms of the application of particular teachings and systems of ethics to key contemporary development concerns. This may be in terms of establishing alternatives to secular materialist development strategies that are considered to be more suitable for followers of that tradition (e.g. Islamic economics or Catholic schools) or broader critiques of mainstream development goals that focus on their location within a western capitalist system where economic growth is the primary desired outcome (for example, we see this type of critique in the religious engagement in the Jubilee 2000 campaign).

In the context of the recent 'turn to religion' in development studies, policy and practice, mainstream development actors are increasingly interested in finding out about the contributions of religious traditions to 'development'. The aim of this chapter is to better understand how we might identify religious approaches to 'development' and the significance of these for mainstream donor-led development. In particular, it aims to address the following questions: *How might members of religious traditions approach 'development'? In what ways does religion influence visions of development as well as the policies and strategies, to reach those visions? Are these religious approaches significantly different from secular approaches?*

In addressing these questions, the chapter is divided into two main sections. First, we will examine several issues that emerge from the fact that within all religious traditions there are a diversity of voices and perspectives; they are *inherently heterogeneous*:

- How are different religions, or denominations within religions, organized such that they appear to speak with 'one voice'?
- To what extent does this 'official' or 'organized' religion give a good account of what people actual believe in and practise?
- To what extent are the ways in which different 'outsiders' (e.g. the academy, policy makers and the media) categorize religious diversity – as along a spectrum from 'liberal' to 'conservative'/'fundamentalist' – analytically helpful in identifying distinct styles of religiosity? Or is such categorization instead guilty of reinforcing colonialist discourses about what a religion should be?

These questions raise additional questions about how we actually find out about religious approaches to development, but also challenge our ability to be able to identify – in a clear, concise and predictable manner – specific religious approaches to key development issues and concerns. Instead, within each religion we are likely to find a range of approaches and perspectives to development problems.

In the second section of the chapter we will examine some of the ways in which adherents of three of the major 'world religions' (Islam, Hinduism and Christianity) have traditionally viewed wealth and poverty, and how their views might consequently shape their understanding of the purpose and desirability of economic development. While there are a broad range of other contemporary development issues that could be selected, as we saw in Chapter 2, in many ways economic development is the most basic, and disputes over the extent to which it should take priority or the shape it should take underpin many disagreements between different development actors across the globe, secular and religious. It is not possible in one chapter to examine more than a few religious traditions, and since most research in this area has been carried out with respect to Christianity and Islam (the two most popular religions globally), it is important to include them here. This also then provides the rationale for exploring Hindu attitudes, since Hinduism (the third largest religion globally) has received

much less attention in recent studies yet is the tradition 'followed by' an estimated 870 million people across the globe (see Table 1.1). Finally, these three
traditions provide interesting contrasts in terms of their current geopolitical
significance and historical relations with the West.

Issues emerging from the heterogeneity of religions

Organizational considerations: religions speaking with one voice, but whose voice?

An examination of the different ways in which religions are organized is important to unpacking how they interact with formal donor-driven development
processes as well as evaluating the nature of the claims made about their
approaches to key development concerns. In Chapter 3 we discussed the way
in which it is typical to think in terms of discrete 'world religions' (Christianity,
Islam, Hinduism, Sikhism, Buddhism, etc.), even if these can be shown to be a
western and relatively recent invention. Within these 'world religions', however,
there are usually different denominations, each attempting to represent the 'true'
or most authentic version of that tradition (e.g. the split between Sunni and Shia
Islam, or between Theravada and Mahayana Buddhism). These denominations
within religions have established hierarchical structures (operating variously at
the international, regional, national and/or local levels) that not only attempt to
influence the individual believer but may also have the organizational and
strategic capacity to influence donors, governments and multilateral bodies in
the Global North, either directly or indirectly. Representatives of these can often
seem to outsiders (as well as insiders) to represent the true or authentic voice
within that particular denomination, or even within the religious tradition as a
whole. Powerful coalitions within particular 'world religions' may even have
the capacity to establish their own multilateral bodies and financial institutions
that reflect their particular religious background (e.g. the Organization of
Islamic Cooperation (OIC) and the Islamic Development Bank (IDB) – see
pp. 99–100; Akbarzadeh and Connor 2008).

However, the processes by which certain groups or voices come to represent
religious constituencies often involve the promotion of the interests of some
and the suppression of those of others. They may use force or coercion, direct
(e.g. violence or legislation) or indirect (e.g. pressure to conform to broader
family or community norms), to enforce their particular version of the denomination or religion. An important part of the process of becoming a 'world
religion', or a denomination within a 'world religion', is establishing centralized
authority that can make judgements on matters of textual interpretation and
other theological issues. This has been one way in which 'world religions' and
their denominations have sought to maintain themselves as unified entities with
a clear idea of what they stand for and what belonging to them entails (e.g. the
Vatican for Catholicism).

Nonetheless, while these 'official versions' of religious traditions seek to establish authentic belief and practice, the ways in which people actually live out their religion also reflects local context and culture. Moreover, while 'official versions' of religious traditions may increasingly seek to establish a religious realm that is demarcated from other aspects of life, the ways in which people actually experience and practise their religion may not reflect such clear processes of differentiation.

The distinction between 'lived' and 'official' religion

The most obvious way to find out about the attitudes held within different religions about key development concerns might seem to be consulting religious leaders, who propound the formal teachings of the tradition, or going first-hand to the actual religious texts or theological treatises produced within different religions. However, this would result in rather a narrow set of views that might not be a true reflection of what people actually believe in and practise. As Beckford writes, to focus on 'official' or 'organized' religion and its institutions is to miss other important aspects of religious life:

> '[O]rganised religion', in the form of, for example, activities conducted in denominations, synagogues, fraternities and mosques is not the entire picture. The label of 'religion' is applied to a vast range of social arrangements and practices partially or wholly outside the sphere of religious organisations. Moreover, the study of religious organisations runs the risk of implying that all participants in these organisations think, feel and act alike.
>
> (2003: 27–28)

In contrast, therefore, to official or organized religion (i.e. that reflected in the 'world religions'), we must also consider what has been called 'lived religion' or 'popular religion' in order to get closer to what people actually do and believe in. Perhaps not surprisingly, as Box 4.1 illustrates, within the 'world religions paradigm' there is also a tendency to view this 'popular religion' or 'lived religion' (sometimes also called 'indigenous religion') as a corrupt form of a true or authentic style of religiosity. In attempting to arrive at an understanding of how people actually live out their religion, and hence to get a clearer idea of its relevance for different development concerns, it is important to reflect on the gulf that may exist between 'lived religion' or 'popular religion' and the ideal or official version of the tradition that is promoted by the religious hierarchy, that can be 'extracted' from religious texts or that is based upon formal theology. The religious hierarchy, typically consisting of male religious leaders, may be out of touch with what people actually do, or may for historical, political, sexist or theological reasons wish to stick rigidly to certain 'official' teachings. Neither may the consultation of religious texts or theological traditions tell us what we want to know. People often do not exactly follow the ideals of their religious traditions, and the texts may not reflect how people actually live their religion. Moreover, religious texts are open to multiple and even conflicting interpretations.

Box 4.1

Village Hinduism

With respect to Hinduism, the gulf between the lived and the textual tradition reflects a distinction between so-called village, popular or vernacular Hinduism and 'Brahmanical' Hinduism. Whereas the former relates to what the majority of people actually do, the latter relates to a version of Hinduism that focuses upon the Sanskrit texts (the Vedas, the Upanishads, the Puranas, etc.) and the beliefs and practices of the priestly Brahmin caste. Studies of Hinduism that emerged during the British colonial period tended to view the Brahmanical religion as the authentic version of Hinduism and saw 'village Hinduism' as an aberration or corruption of the pure textual tradition. Therefore, linked to this consideration of the gap between lived religion and the textual tradition is the broader issue of the extent to which the very notion of Hinduism is a colonial construction.

Source: adapted from Tomalin (2009b).

This is not to say that 'texts' tell us nothing about how people practise religion. Rather, it is to say that they cannot be taken as a straightforward reflection or indication of the nature of religious activity and its relations to the social world. Religious traditions are not homogeneous, and local cultural variations in religious belief and practice have an impact on the diverse ways that they are played out in different locations. For this reason, it may be inappropriate for an 'outsider' to attempt to read religious attitudes from the texts, when these only have significance in the light of an understanding about how they are actually lived by people. The difficulty of extracting particular religious views on development (or any other topic) purely from talking to religious leaders or consulting religious texts and theologies points towards the value of empirical or ethnographic research that captures the various ways in which people approach their religious traditions in practice.

'Liberal' or 'fundamentalist'? 'Outsider' classifications to describe and evaluate different 'theological' or 'insider' positions

The interpretation of religious texts and traditions can result in a range of positions within particular religions, and this has attracted the use of different descriptive labels by 'outsiders' (academics, policy makers, the media, etc.). These terms are also used by members of religions themselves, either in self-reference or with respect to others. A distinction is often made between 'conservative' and 'liberal' religion. Adherents of conservative religion typically emphasize that the central religious texts either were revealed word for word to prophets or sages, or were divinely inspired. Therefore, the Qur'an for Muslims, the Torah for Jews, the Bible for Christians and the Vedas for Hindus are considered to be the word of God, telling believers the exact nature of their relationship to the divine and providing guidance as to how they should live.

Conservative Christians, for instance, believe that the Bible is inerrant and refuse to reinterpret it in the light of modern problems and concerns. This is a particular development following the Protestant Reformation, when the Bible came to be considered as a text that could directly form a link between humans and the divine without the need for priests or the Catholic Church to act as intermediaries.

However, many traditions rely not only on their foundational religious texts but also upon the theologies, philosophies or systems of law that develop in relation to them. For instance, as Box 4.2 illustrates, Shari'ah law in Islam, which is supposed to govern all aspects of a Muslim's life, primarily draws upon two main sources: the Qur'an (the foundational Muslim holy book, which is believed to be the direct word of God) and the Sunnah (the sayings and conduct of the prophet Muhammad, as found in the Hadith). However, the process of interpreting these two sources is known as *fiqh* (Islamic jurisprudence), and whereas the Qur'an and the Hadith are infallible, *fiqh* may change according to context. Thus, religious texts are often inconclusive in their meaning, and different interpretations or theologies have been developed (which may vary over time and between interpreters) in order to make sense of these texts and to develop a more coherent sense of the nature of God's intentions. For instance, Catholic theology and its resultant social teachings have been shaped by the Bible but also by the writings of different theologians throughout its history and the influence of the *magisterium*, the teaching authority of the Church that resides in the Pope and his bishops, with its centre in the Vatican in Rome. Thus, while Catholicism is typically considered to be 'conservative', it draws its authority from its traditions of theology and liturgy as much as from the Bible.

While the distinction between conservative and liberal does not directly refer to the way that traditions relate to social and political issues, conservative religions are more likely to hold conservative views on gender roles and sexuality, for instance. The Catholic attitude towards contraception and abortion is an example of the way in which a religious tradition may adhere to teachings about the nature of the human being and the role of sexual intercourse, despite liberal critiques and counter-viewpoints widely held in many societies.

Liberal religionists, by contrast, do not believe in the literal word of religious texts and instead consider that although they reflect religious truths, religions need to change over time. Moreover, the fact that they were written down by humans, at particular points in time, means that they are likely to be coloured by the preferences of their authors as well as the social context. As Box 4.3 illustrates, a good example of the 'liberal religious' approach is the emergence of various 'feminist theologies' since the 1960s. According to the liberal view, which includes feminist theology, while religious texts are important, their truth needs to be discovered through interpretation, which can both uncover their original significance and make them relevant to the social and political context of the current day.

Box 4.2

The Shari'ah (literally: 'the path leading to the watering place')

Kroessin explains that

> the *Shari'ah* deals with many aspects of day-to-day life, including economics, banking, business law, contract law, sexuality, social issues and dietary laws. . . . Some Islamic scholars accept *Shari'ah* as the body of precedent and legal theory, whilst other scholars view *Shari'ah* as a changing body of law.
>
> (2008: 19)

Specifically, Shari'ah law is divided into two main dimensions. The first is the 'acts of worship', *al-ibadat* (these are also called the 'five pillars' of Islam):

1 ritual purification (*wudu*);
2 prayers (*salah*);
3 fasts (*sawm*);
4 charity (*zakat*);
5 pilgrimage to Makkah (*hajj*).

The second is concerned with 'human interaction', *al-mu'amalat*:

1 financial transactions;
2 endowments;
3 laws of inheritance;
4 marriage, divorce and child care;
5 foods and drinks (including ritual slaughtering and hunting);
6 punishments (*hudood*);
7 warfare and peace;
8 judicial matters (including witnesses and forms of evidence).

Islamic legal theory has established five categories by which acts are labelled as either:

1 obligatory (*wajib*);
2 recommended (*mandub*);
3 permissible (*mubah*);
4 prohibited (*haram*);
5 repugnant (*makruh*).

Source: adapted from Kroessin (2008).

Box 4.3

Feminist theologies

A good example of the 'liberal religious' approach is the emergence of various 'feminist theologies' since the 1960s. With the rise of the women's movement, many women became increasingly aware that their religious traditions were deeply patriarchal, typically portraying them as secondary to men and defining their roles in terms of biology (Daly 1973). However, rather than abandoning religion because of this, many decided to remain within their religious traditions but to subject them to feminist reinterpretations (Ruether 1983; Mernissi 1975; Ahmed 1992; Singh 1993). They argued that religious texts and traditions had been influenced by patriarchal values, rendering them essentially at odds with the 'true' message of the religion. This could be discovered through various methods of feminist analysis, including feminist hermeneutics that aimed to reveal the 'herstory' that had become masked by the 'history' in the majority of religions (Schussler-Fiorenza 1983).

In addition to the distinction between conservative and liberal religion, another classification that is increasingly employed is 'fundamentalist'. Sometimes the term is used interchangeably with 'conservative', as they both draw attention to the inerrancy of scripture and tend to be socially conservative, for instance in their views about women. However, generally the term 'fundamentalist' is used to refer to ultra-conservatism, although the exact meaning may differ according to the context in which it is being used. The term first emerged in North America in the nineteenth century in self-reference by a movement within the conservative Southern Baptist Church, which aimed go back to the 'funda-mentals' (i.e. the Bible) as the basis of tradition (Hawley and Proudfoot 1994), but is now more widely applied to expressions within all religions that are viewed as having certain 'family resemblances' that mark them out as fundamentalist (Marty and Appleby 1991). Box 4.4 outlines these 'family resemblances' according to the findings of the 'Fundamentalism Project' (1987–1995).

Box 4.4

Fundamentalism

The multi-authored 'Fundamentalism Project' was funded by the American Academy of Arts and Sciences (1987–1995) and was directed by Martin E. Marty and R. Scott Appleby. Five volumes were produced. It is observed that there are a series of 'family resemblances' between different examples of fundamentalism. These include (Marty and Appleby 1991: 817–833) the following:

- Religious idealism serves as the basis for personal and communal identity.
- Fundamentalists understand truth to be revealed and unified.
- Fundamentalism is intentionally scandalous.
- Extremist values and behaviours seek to separate true believers from outsiders.
- Fundamentalists envision themselves as part of a cosmic struggle.
- They seize on historical moments and reinterpret them in light of this cosmic struggle.
- They demonize their opposition and are reactionary.
- They name and demonize their enemies.

- Fundamentalists are selective in what parts of their tradition and heritage they stress.
- They are led by males.
- They envy modernist cultural hegemony and try to overturn the distribution of power.

Marty and Appleby write:

[F]undamentalism has appeared as a tendency, a habit of mind, found within religious communities and paradigmatically embodied in certain representative individuals and movements, which manifests itself as a strategy, or set of strategies, by which beleaguered believers attempt to preserve their distinctive identity as a people or group. Feeling this identity to be at risk in the contemporary era, they fortify it by a selective retrieval of doctrines, beliefs, and practices from a sacred past. These retrieved 'fundamentals' are refined, modified, and sanctioned in a spirit of shrewd pragmatism: they are to serve as a bulwark against the encroachment of outsiders who threaten to draw the believers into a syncretistic, areligious, or irreligious cultural milieu.

(1991: 835)

Many scholars are nonetheless uneasy about the use of the term since those it describes typically do not use it. The postcolonial studies scholar Edward Said (1993) considered terms such as 'fundamentalism' and 'terrorism' to be modern-day colonialist responses to types of religion that do not fit its mould (see also Munson 2005: 338). Box 4.5 discusses Tariq Ramadan's argument that the term fundamentalist is used to denote 'bad Muslims', whereas 'good Muslims' are seen as 'those who are invisible, or look just like us' (2010; see also Mandani 2004).

Box 4.5

'Good Muslim, Bad Muslim'

Tariq Ramadan (2010) argues that the search for 'moderate Muslims' after 9/11 reproduces colonial categories where

'Good' Muslims were those who either collaborated with the colonial enterprise or accepted the values and customs of the dominant power. The rest, the 'bad' Muslims, those who 'resisted' religiously, culturally or politically, were systematically denigrated, dismissed as the 'other' and repressed as a 'danger'.

Today, 'bad Muslims' are labelled as fundamentalist, extremist and Islamist, whereas good or moderate Muslims are seen as 'those who are invisible, or look just like us, who support us, or even as those who have accepted the terms of their subjection' (ibid.). Ramadan argues that while Islam across the globe adheres to certain 'fundamental principles' (e.g. the 'five pillars'), liberal or moderate approaches are actually part of its history, involving interpretations that are flexible to local contexts and changing times. From a western perspective, obligations such as praying five times a day or adhering to certain dress codes seem to lack moderation, but this does not mean that Islam lacks flexibility in other ways. The western way of seeing religion means that

> moderate Muslims are those who adopt no distinctive dress, who consume alcohol and practise their religion 'as we do ours' – that is, not really, or by making it invisible in the public sphere. But our histories, cultures and reference points are not identical; the notion of moderation has to be studied from within each system of reference. It cannot be imposed from outside.
>
> (ibid.)

* * *

This first section of the chapter has aimed to examine the implications of a series of issues for finding out about and evaluating religious attitudes in the context of understanding their implications for development processes and outcomes. Each of these issues arises from the fact that all religious traditions are *inherently heterogeneous*, and two main sets of observations can be summarized. First, it is important to be aware of the gulf that may exist between 'official' and 'lived' religion, and to adopt a range of research methods: talking to religious leaders and examining interpretations of religious texts as well as carrying out ethnographic research in order to better understand how people actually 'live' their religion. Second, the terminology that is typically used to describe religious diversity is also evaluative and can reflect a western bias. Labels such as fundamentalist, extremist, conservative or liberal may be useful in describing particular manifestations of religious activity, but the fact that they may also be underpinned by a politics that serves the interests of those using them should not be overlooked.

In the second section we will examine some of the ways in which members of three of the major 'world religions' (Islam, Hinduism and Christianity) have traditionally viewed wealth and poverty, and how their outlooks might shape their approach to economic development. This exploration will involve an examination of traditional religious teachings and attitudes, as well as the ways in which these have been applied more recently to contemporary international development discourse.

Religious approaches to wealth, poverty and economic development

The following discussion mainly draws upon a series of literature reviews that were prepared by researchers working on a five-year research programme which was based at Birmingham University in the United Kingdom and funded by DFID, the United Kingdom's Department for International Development. These examined the links between religions and development (for Islam, Kroessin 2007, Zaman 2008; for Hinduism, Tomalin 2009b; and for Christianity, Kim 2007). Each review summarizes recent material produced by relevant religious organizations and academic interpreters to provide background material on the understandings of 'development' that arise out of the core beliefs and values of each faith tradition. I shall begin our discussion of each tradition with some brief comments about their significant features, including (in the case of Islam and Hinduism) the history of their relationship with the West. Then, for each tradition we will examine attitudes towards wealth and poverty, including definitions of poverty, its causes and significance; motivations and mechanisms for helping the poor; and approaches towards economic development.

Islam

> As the lowest common denominator for all Muslim denominations or sects, we can assert that Islam is regarded as a monotheistic religion based upon the teachings of the Prophet Muhammad [570–632 CE], with an estimated 1.7 billion adherents, known as Muslims. Historically Islam spread from the Arabian peninsula and today is present as a majority religion in many countries from West Africa to South East Asia, although there are also sizeable Muslim minorities in Northern America and Western Europe. Only about 20 per cent of Muslims live in Arab countries (Indonesia is the most populous Muslim country) and only approximately 10 per cent speak Arabic as their mother tongue. Linguistically, Islam means submission, referring to the total surrender of one's self to God (in Arabic *Allâh*).
>
> (Kroessin 2008: 10)

Before we examine some Islamic attitudes towards wealth, poverty and economic development, it is helpful to briefly explore the different strands of literature about Islam and Muslim contexts that could inform this inquiry. During the eighteenth and nineteenth centuries the study of Islam in the West reflected the interests of colonial administrators as well as a sense of rivalry between the West and the Muslim world. However, a new paradigm emerged in the twentieth century in the form of 'Islamic studies', generated by both western and Muslim scholars. As Kroessin writes, this was essentially a social science discipline and should not to be confused with the understanding of 'Islamic studies' in some Muslim contexts, as an 'umbrella term for virtually all the social science disciplines, since Islam sees itself as offering guidance on all human affairs' (2008: 9). This new band of scholars of Islam sought to avoid the construction of 'Islam as a distinct homogeneous and timeless entity that is essentially informed only by its normative texts (the *Qu'ran* and the *hadith*), in the way orientalists used to' (Kroessin 2008: 8). 'Orientalism', a term coined by Edward Said, can be defined as 'a Western style for dominating, restructuring, and having authority over the Orient' (Said 1979: 3). Particularly since

9/11, however, there has been a renewed interest in the Muslim world globally. Some of this reflects a rekindling of orientalist approaches in western scholarship, particularly within many post-9/11 studies of Islam that are focused on the war against Islamic fundamentalism and terrorism (e.g. Burr and Collins 2006).

In terms of the study of Islam, also of interest are 'insider' writings from within Islamic contexts reflecting 'Islamic revivalism' and the emergence of what has been called 'political Islam', associated with writers such as Sayyid Abul A'la Maududi (d. 1979) in South Asia and Sayyid Qutb (d. 1966) in the Middle East. In modern Islamic revivalism there is a lack of separation between religion and politics, based on the view that Islam can only be properly adhered to within an Islamic State where Shari'ah law is followed. Closely linked to these trends has been the emergence of what has been called 'transnational Islam' or 'pan-Islamism' (see below, p. 99, Box 4.7). Of direct relevance to the discussion here, there is a growing literature on Islamic economics, which can be located within Islamic teachings and practices over centuries, but more recently has been part of Islamic revivalism (see Zaman 2008).

Attitudes towards wealth

Wilson (1997: 117) points out that over 1,400 of the 6,226 verses in the Qur'an deal with economic matters, 'many of them in a highly specific manner, particularly in relation to trade, finance and inheritance' (Kroessin 2008: 44). Despite this emphasis on economics, as Kroessin notes, 'the human needs identified in Islam are not exclusively material and cannot be captured in monetary terms alone – the Islamic understanding of poverty transcends the physical world, with its emphasis on spiritual well-being' (ibid.: 47). While the literature on Islamic approaches to development has largely focused on economic issues, the emphasis in revivalist Islamic social teachings is not just on materialism but also on compliance with Islamic Shari'ah law in all spheres of human existence (Kroessin 2011). For instance, while Muslims are allowed to own property, they are banned from 'speculating, gambling, hoarding and destructive competition. Included in this last group are activities such as participating in insurance or capital markets (gambling) or property dealing (speculation)' (Kroessin 2008: 49). Crucially, according to Shari'ah law, Muslims may not earn any interest on money they deposit in the bank, which has encouraged the development of an Islamic banking system across the Muslim world. The development of Islamic economics has been seen as important in strengthening Islamic revival in Islamic states, but also for Islamic authority to be established in areas where western influences dominate (ibid.: 50).

Defining poverty in terms of human needs

The modern development organization Islamic Relief (an international aid and development charity founded in the United Kingdom in 1984) has applied

Islamic teachings about 'human needs' to underpin its understanding of poverty. Poverty is the result of human failure to follow Shari'ah law, but in Islam is understood as relating to more than economic poverty. While poverty in main-stream development discourse has traditionally focused on economic poverty, more recently this has shifted to a multidimensional approach, and it is here that Islamic Relief locates the Islamic view. As Box 4.6 demonstrates, particular efforts are made by Islamic Relief to engage with modern development dis-course, and the traditional Islamic view on poverty is laid down beside 'secular' thinking on the topic.

Helping the poor: poverty alleviation

Islamic texts stress the importance of giving to the poor, and this had become mandatory in the form of *zakat* (one of the 'five pillars' of Islam) by the time Muhammad established the first Islamic state in 622 CE. This is the primary form of charitable giving in Islam and is to be given at the end of every year by each Muslim, calculated at 2.5 per cent of any disposable wealth above a minimum amount. It may only be used to benefit Muslims, and Shia Muslims do not believe that it should be collected and administered by the state:

> [*Zakat*] expenditures are only for the poor and the needy, and for those employed to collect [*zakat*] and for bringing hearts together and for freeing

---

Box 4.6

Islamic views on poverty

According to the faith-based organization Islamic Relief:

> Comparing the various interpretations of poverty with Islamic guidance on justice, development and support, Islamic Relief understands poverty as a multidimensional phenomenon, with a special focus on capability deprivation. Poverty encompasses not only material deprivation (measured by income or consumption), but also forms of deprivation such as unemployment, ill health, lack of education, vulnerability, powerlessness, and social exclusion. In Islam, poverty is defined by five groups of activities and things which make up the human needs: (a) religion, (b) physical self, (c) intellect or knowledge, (d) offspring and family, and (e) wealth. The fulfilment of these needs is considered one of the basic goals of Islam.[1]
>
> These needs define the foundations for good individual and social life, and are classified into three types: (1) necessities (*dharuriyyat*), (2) convenience (*hajiat*), and (3) refinements (*kamaliat*). Necessities consist of all activities and things that are essential to preserve these five needs . . . at the barest minimum . . . at this level, one has enough to live but not necessarily enough to be in some comfort.
>
> (Islamic Relief 2008: 11)

People who are unable to meet the necessities are poor. People who are unable even to come halfway are considered destitute.

captives and for those in debt and in the way of Allah and for the traveller – an obligation imposed by God and God is Knowing and Wise.

(Qur'an Surah 9, verse 60, cited by Kroessin 2008: 47)

Kroessin (2008: 47–48) translates this passage into 'modern public finance terms' as relating to the following expenditure headings:

- poverty reduction;
- administrative overheads for civil servants dealing with public welfare;
- peace-building and community cohesion;
- promotion of freedom, human rights and civil liberties;
- personal insolvency settlements;
- security and defence;
- homeless, refugees and migrants.

In addition to *zakat*, which is often given via an intermediary, including the state (e.g. in Sudan, Malaysia and Pakistan), there are other mechanisms for direct giving. For instance, during the yearly month of fasting, Ramadan, Muslims are expected to feed the destitute while fasting themselves; they are expected to make a contribution called *zakat al-fitr* at the end of Ramadan, comprising 'one saa of food . . . equivalent to four maad . . . [where a] maad is the amount that can be scooped up when one puts their hands together'[2] or its monetary equivalent; and during the *hajj*, whether on pilgrimage or not, they are expected to perform *qurbani*, the sacrifice of a cow, goat or camel to feed the needy (Kroessin 2008: 48). In addition to *zakat* we also find *sadaqah* (voluntary charity), which is an individual devotion where charity is given directly to a beneficiary.

However, in the modern era most Muslims do not live in Islamic states, and therefore charities and development organizations play a role in collecting *zakat* and *sadaqah* in Muslim (and increasingly non-Muslim) contexts, as well as carrying out *zakat al-fitr* and *qurbani* on behalf of devotees wishing to help other Muslims (Plates 4.1 and 4.2). Most modern Islamic development organizations consider that they are carrying on these traditions (Benthall and Bellion-Jourdan 2003), and, as Kroessin writes, '[f]undraising efforts based on individual donations appeal to individual Muslims to discharge their Islamic duty for annual *zakat* payment on the basis of verses from the *Qur'an* and *ahadith*' (2008: 53).

Another means of charity is the establishment of a 'pious foundation' by endowment (*waqf*). As Kroessin explains, the '[u]se of funds generated by a *waqf* can be utilized for all aspects of charity work, including construction of mosques, shelters, orphanages, refugee camps and hospitals, historically evidenced by many endowments remaining active today in Turkey and the Middle East' (2008: 48–49).

Plate 4.1 *Islamic Relief* qurbani *programme 2011, Niger*

Source: Islamic Relief UK.

People queuing to receive meat from *qurbani*.

Islamic economics

A distinct field of Islamic economics emerged as a new discipline in the early twentieth century when 'a desire for an Islamization of deeply secular disciplines like economics became a priority during post-colonial struggles in the Muslim world' (Kroessin 2008: 50). As Zaman (in his survey of the literature on Islamic economics) writes, the emergence from colonialism placed many Muslims across the globe in the position of needing to 'find a suitable compromise between the demands of modernity and the demands of Islam' (2008: 20) as they entered a postcolonial globalized world dominated by capitalism and western institutions. Emerging from the needs of the Islamic revivalist liberation movements, which asserted that there was a better way for Muslims to live than according to western norms, Islamic economics was born.

There have been many Muslim thinkers in this field, but Zaman (2008: 18) identifies two in particular, Mohammad Baqir Al-Sadr (*Our Economy*, 2000 [1961]) and Sayyid Abul A'la Maududi (e.g. *Economic System of Islam*, 1970). Zaman (2008) outlines in some detail the fundamentals of an Islamic economic system. While a spectrum of positions have been adopted by different thinkers,

Plate 4.2 *Islamic Relief* qurbani *programme 2011, Lebanon*

Source: Islamic Relief UK.

Woman receiving meat from *qurbani*.

he demonstrates that 'there is widespread agreement among Muslims with the theses of Maududi and Baqir Al-Sadr that the moral imperatives embedded in an Islamic economic system differentiate it from both capitalism and communism' (ibid.: 20). Whereas in capitalism, for instance, economic growth is an end in itself and wealth is used as a yardstick by which to measure development, this focus has been criticized by Islamic economists. For instance, as Maududi (1970) writes,

> economic problems, which form a part of the larger problems of human existence, have been separated out of the collectivity and considered in isolation. This tendency eventually grew to the extent that economic problem[s] became the central problem[s] of human existence. This second mistake was even larger than the first.
>
> (cited in Zaman 2008: 27)

The main contrast, Zaman suggests, between Islamic models and neo-classical economics is that Islam views human beings as capable of being transformed from 'followers of base desires to people concerned with achieving higher goals' (2008: 28). Whereas Islam is considered to promote cooperation between individuals 'who can be motivated to be generous', neo-classical economics is instead based on 'cut-throat competition between selfish individuals' (ibid.: 30). According to Zaman,

> [t]his basis for solving the economic problems of humankind differs radically from any currently on the conventional economic menu (e.g. the World Bank website). Islam offers a vision of a society based on cooperation and community harmony, people who take care of each other in times of need, and an economic system to match. Historically, Islamic civilization has taken much better care of its disadvantaged and poor, as well as minorities (by offering them religious and cultural freedom), than others, including current European civilization. Is it not worth struggling to create such a world? That is what Islamic economics is about.
>
> (ibid.: 80)

Economic development in Islamic contexts

This new field of Islamic economics, emerging from the revivalist movements of the twentieth century, has had a direct influence on approaches to economic development in Islamic contexts. The purpose of economic policy within Islamic contexts is not to stimulate economic growth for its own sake but instead to seek to rebalance economic inequalities through redistributive strategies such as *zakat* and also through preventing individuals from accumulating massive wealth via interest, in compliance with Shari'ah law.

By the 1970s, factors such as the creation of OPEC (Organization of the Petroleum Exporting Countries), which placed power and wealth in Muslim hands, meant that there was a need to create Islamically permissible alternatives to the dominant western financial markets and institutions (Zaman 2008: 20). In order for Islamic economics to be viable, Islamic financial institutions were established. Box 4.7 outlines the main purpose of the largest of these, the Islamic Development Bank (IDB), which was set up by the Organization of the Islamic Conference (OIC) in 1973. (In 2011 the name was changed to the Organization of Islamic Cooperation.)

Box 4.7

The Islamic Development Bank

The website of the IDB states that '[t]he purpose of the Bank is to foster the economic development and social progress of member countries and Muslim communities individually as well as jointly in accordance with the principles of Shari'ah i.e., Islamic Law'.[3]

It has established a 'fund for poverty reduction' using a 'south–south' approach with specific Islamic characteristics,[4] and 'extends loans to its member countries for the financing of infrastructural and agricultural projects such as roads, canals, dams, schools, hospitals, housing, rural development, etc. both in the public and private sectors'.[5] However, 'such loans, in conformity with Shariah, are interest-free and the Bank recovers its administrative expenses by levying a service fee'.[6]

The creation of the IDB was part of broader shifts in the Middle East towards embracing 'pan-Islamism', 'the movement which aims to unite the diversified Muslims on the basis of their common religion' (Rauf 2007: 21), including the desire to create an Islamic state or states. Tensions in the Middle East between the Saudi leadership and secular pan-Arabism, led by Gamal Abdul Nasser in Egypt, as well as the Saudis' fear of Gaddafi's rise to power in 1969, led to the establishment of a series of organizations. As Kroessin writes, '[b]y embracing pan-Islamism, Saudi King Faisal could counter the idea of pan-Arab loyalty centred around Egypt with a larger transnational loyalty centred on Saudi Arabia' (2008: 27). The Muslim World League was set up by Muslim religious figures from twenty-two countries in 1962 in Saudi Arabia to serve as a representative body for Muslims, though it is funded and dominated by Sunni Saudi Arabia (ibid.: 28). In 1969 the Organization of the Islamic Conference (OIC) was set up after the Arab–Israeli war of 1967 as an overall body for pan-Islamic political leadership, with its headquarters in Jeddah, Saudi Arabia. It represents fifty-seven states across the globe, most of them Islamic. After the United Nations it is the next-largest international organization and has a permanent delegation to the United Nations. No such pan-religious organizations exist on the same scale within other religious traditions.

However, Muslims across the globe live in a range of different contexts, from Islamic (e.g. Iran – Niblock and Wilson 1999) and partially Islamic states (e.g. Malaysia – Ariff 1998; Indonesia – Timberg 2000; and Pakistan – Ayub 2007), to western-style secular democracies. As Zaman writes:

> In states where Muslims have had political power and live in sufficient numbers, fully or partially Islamic systems have resulted. In other situations,

> Muslims have used private and market-oriented means rather than government policies to create economic institutions in conformity with Islamic law.
>
> (2008: 21)

Private-sector solutions include Islamic mortgage companies, mutual funds and day-to-day banking (ibid.: 23–24). However, as Kroessin points out, some critics have argued that 'Islamic financial products merely aim to replicate in Islamic form the functions of contemporary financial instruments, markets and institutions, and have therefore, arguably failed to serve the social objectives of Islamic law' (2008: 52).

Hinduism

In thinking through Hindu approaches to wealth, poverty and economic development, we are faced with the task of unpicking thousands of years of history and numerous texts written in Sanskrit as well as vernacular languages. Hinduism is normally considered to have its origins in the Indus Valley civilization (*c*.2500 BCE to 1500 BCE) and the later Aryan culture, which developed during the second millennium BCE. The exact relationship between these two cultures and the ways in which they fed into the development of what we now call the Hindu tradition has been much debated. Whichever view one takes, Flood argues that 'Hinduism might be regarded as the development over the next 2,000 years of Aryan culture, interacting with non-Aryan or Dravidian and tribal cultures, though it is Aryan culture which has provided the "master narrative", absorbing and controlling other discourses' (1996: 23).

Moreover, as was discussed in Chapter 3, criticisms that the idea of Hinduism as a 'world religion' was a colonial and Christian-influenced invention are persuasive, drawing attention to the difficulty of transposing the 'world religions' model to all corners of the globe. For instance, Hinduism had no founder (unlike Islam or Christianity) and even today is constituted by a diverse array of practices and teachings, associated with a pantheon of pan-Indian male and female deities, but also thousands of local deities, as well as distinct philosophical traditions that outline different positions on the relationship between humans, the divine and the material world. Nonetheless, there are certain common threads running through these diverse expressions of religiosity (such as caste, *karma* and rebirth), and today the notion of being a 'Hindu' is a concept that many people in South Asia and different diasporas employ to describe their religious identity.

As with the study of Islam, the beginnings of scholarly interest in Hinduism in Europe took off significantly in the eighteenth century, although there had been some earlier interest in accounts of travellers and missionaries. The British administration in India required information about the religious practices of the subcontinent. But we also find a passionate band of 'orientalist' scholars, who valorized and romanticized the grand Indian traditions reflected in the ancient texts (the Vedas and the Upanishads in particular), and who generated a vast body of literature that helped to shape Hinduism as a 'world religion' that could be the object of study in western institutions (Klostermaier 1994: 17ff.). Both reflecting and reinforcing the emergence of Hinduism as a 'world religion', a number of Hindu reform movements came about from the late nineteenth century onwards (including the Brahmo Samaj and the Arya Samaj), with an interest in defining the boundaries of Hinduism vis-à-vis other religious traditions in India. These reform movements have as their legacy the contemporary Hindu nationalist movement, as outlined in Box 4.8, which is made up of a number of religio-political organizations that aim to promote and protect what they consider to be an authentic Hindu identity for all Indians, thereby marginalizing other religions in India and even creating violence and hostility towards members of those religions. Many of these Hindu nationalist organizations have been very successful in establishing social welfare and development organizations across the country, causing some to express concern that there has been a 'Hinduization of civil society in India', underpinned by divisive sectarian forces. As Jodkha and Bora argue, awareness of this is

> certainly relevant to a consideration of faith-based activity in India, where accusations of the 'Hinduization' of civil society, in particular, can mean that individuals and organizations often avoid overt identification with religious values and rhetoric in order that they are not aligned with the Hindu Right.
>
> (2009: 5)

Attitudes towards wealth

As with Islamic views, from a Hindu perspective other things, apart from economic wealth, are significant for a person's well-being and 'development'.

Box 4.8

Hindu nationalism

Hindu nationalism has its roots in the emergence of religious identities in colonial India, as well as the Hindu reform movements and nationalist politics of the pre-independence era. In all communities – Sikh, Hindu and Muslim – people began organizing along religious lines. For Hindus, this meant 'delineating a broad-based communal identity beyond caste that had not been emphasized strongly before' (Gold 1991: 537). There has emerged from the Hindu nationalist movement not only a strengthening of the idea of Hinduism as a world religion but also the ideology of *Hindutva* ('Hinduness'), the key tenets of which include the idea that the Indian subcontinent is the homeland of Hindus and that Hindus are those who follow the original Indic religion of the Vedas (the most ancient Hindu texts). One of the most important groups propounding this view is the Vishva Hindu Parishad (VHP), founded in 1964 as a religio-cultural organization that attempts to articulate a universal Hinduism (McKean 1996: 102). Other groups, alongside the VHP, that also promote *Hindutva*, such as the Rashtriya Swayamsevak Sangh (RSS, founded in 1925) and the Bharatiya Janata Party (BJP, founded in 1980), are collectively known as the *sangh parivar* (family of organizations).

Source: adapted from Tomalin (2009b).

Within Hinduism a number of attitudes towards the accumulation of wealth are reflected. First, it is a legitimate aim for humans in terms of *artha*, one of the four 'aims of life' (*purusharthas*): *artha* (wealth/material prosperity), *dharma* (duty/rules), *kama* (enjoyment) and *moksha* (spiritual liberation). Second, it is something that prevents people from realizing the highest goal of spiritual liberation (*moksha* – liberation from the cycle of rebirths, *samsara*), since wealth and other trappings of the material world play a part in binding the individual to future rebirths. However, there is also a sense in which material wealth and security can be seen as a sign of religious virtue – that is, an outcome of previously meritorious behaviour. There seems to be a tension here between understanding poverty as a virtue (that is, lesser attachment to the material world means that one is more likely to achieve *moksha*) and understanding wealth as a virtue (that is, because it is a sign that one has done one's *dharma* (duty) well in a previous life). Similar tensions can be found within other 'world religions'.

Poverty and the caste system

Whereas discussions about Islamic attitudes towards poverty focus on how poverty is defined and understood, as well as motivations and mechanisms to deal with it, much literature about Hinduism and poverty tends to focus on its causes, and in particular upon the relationship between poverty and the caste system. Many consider that the caste system in India, which, as Box 4.9 suggests, is supported by the Hindu tradition, continues to play a role in maintaining poverty and inequality. While one cannot claim that poverty is India has been a product of the caste system alone, it has no doubt been an important factor in maintaining socio-economic inequalities in India for at least a couple of millennia. As Klostermaier writes, 'caste has been seen as an essential institution of Hinduism from the very beginning, both by Hindus and outsiders' (1994: 333). The caste system has traditionally sustained a vision of society that is non-egalitarian. It comprises the hierarchical division of society into castes or *jatis*, historically linked to occupation, into which one is born according to actions in a previous life (*karma*). In particular, the caste system and the associated view that women are a lesser rebirth than men clearly present problems for the egalitarian values that drive modern development policies. The social ethics underlying the caste system are not just ideals projected by the Sanskrit texts but are actually manifest within the ordering of society. There is a substantial historical, sociological and anthropological literature on the caste system (e.g. Bayly 1999; Dirks 2001; Dumont 1970; Srinivas 1962), and caste issues have come to shape many aspects of Indian social and political life, having a concrete effect upon social and economic opportunity. This would seem to result in a rather fatalistic view of poverty, which is seen not only as the result of actions in a previous life but also as something that can only be escaped in a future life through 'good behaviour' that results in a better rebirth.

While caste does play a role in locking people into situations of poverty, there have been attempts to circumvent the negative impacts of this social institution.

Box 4.9

Caste and Hinduism

The Hindu tradition has traditionally sustained a vision of society that is hierarchical and non-egalitarian. There is some suggestion in the Vedas, the earliest texts associated with Hinduism, which date from the second millennium BCE, that society was already divided into four classes or *varna*s: the priestly Brahmin class; the warrior or *kshatriya* class; the merchant or *vaishya* class; and the servant or *shudra* class. The top three *varna*s were called 'twice born', a reference to the initiation that boys undergo – the *upanayana* – as being like a second birth. These were traditionally the only *varna*s allowed to hear the Vedas being recited. There is a hymn in the 'Purusha Sukta' of the Rig Veda (the oldest Vedic text) that describes the emergence of these four classes as a cosmic sacrifice performed by Purusha (the 'cosmic man').

The mythical figure of Purusha is often taken as marking the origins of the caste system, although the notion of *varna* ('class') is not actually identical to that of 'caste' (*jati*). *Jati* means 'birth', and while there are only four *varna*s, there are literally thousands of *jati*s. These are endogamous sections of society traditionally associated with different types of occupation and are arranged in terms of relative levels of ritual 'purity': the Brahmins (this can be the name both of a *varna* and of a *jati*) are considered to be at the top and the outcastes or untouchables at the bottom. As Flood points out, while 'the exact historical relationship between *varṇa* and *jāti* is unclear . . . [t]he traditional view is that the *jātis* represent a proliferation of social groups from the *varṇa* system' (1996: 60).

Source: adapted from Tomalin (2009b).

For instance, the Indian state provides job and educational reservations for low castes, and the practice of 'untouchability' – where the lowest castes were discriminated against in almost all areas of life – was outlawed in 1950. Nonetheless, the former untouchable castes still exist and are now known as 'Dalits' (downtrodden). There are strong Dalit movements in contemporary India that continue to fight for Dalit rights and to counter still-existing forms of discrimination (Plate 4.3). As Box 4.10 illustrates, one important strategy has involved conversion to Buddhism or another religion to escape the caste system.

Box 4.10

The Dalit movement and Dr Ambedkar

While Dr Babasaheb Ambedkar (1891–1956) was born as an outcaste or untouchable, he managed to pursue higher education in the United States and the United Kingdom and to become a lawyer and Bahujan political leader, as well as the main author of the Indian constitution after Independence. However, his pursuit of Dalit rights was not confined to politics: he converted to Buddhism in 1956 and then converted an estimated 380,000

Plate 4.3 Dalit rights protest, New Delhi, 1 August 2012

Source: AFP/Getty Images.

Christians are pictured holding crosses during a rally in protest against the National Commission for Scheduled Castes and Scheduled Tribes, which recently rejected the demand for 'reservation' for Dalit Christians and Muslims. Reservation refers to the setting aside of a percentage of educational opportunities and government jobs for Dalits. However, currently these are only available to Hindus, Buddhists and Sikhs.

of his followers to the tradition. This conversion was a symbolic rejection of Hinduism, which was considered, through its support for the caste system, to justify the oppression of outcastes. Moreover, conversion to Buddhism has proved to be an important means of enabling Dalits to enhance their sense of self-worth and their social standing in a society that continues to exercise discrimination against what are now called, for classification purposes, Scheduled Castes and Scheduled Tribes (SC/ST). Studies on Dalit communities indicate consistently low levels of education, health and income compared to those belonging to higher castes. Increasing numbers are converting to other religions, including Buddhism, in order to improve their status (Fitzgerald 1990).

Source: adapted from Tomalin (2009b).

Helping and giving to the poor: poverty alleviation

While Hinduism could be interpreted to support the view that poverty is caused by 'bad' actions in a previous life, and therefore the poor are only getting what they deserve, there is also an important tradition of selfless giving (*dana*) and service (*seva*), for instance to temples and holy men, but also to the poor. These practices are considered to be part of the duty (*dharma*) of each individual and contribute towards positive *karma* that will carry forward to future rebirths. However, these practices must be performed without attachment to personal gain or reward. Often people will donate food or clothing, for instance, directly to those in need as part of this spiritual practice, or will contribute towards the activities undertaken by different Hindu charities and FBOs (see Box 8.2 on p. 209).

The example of Swami Vivekananda, one of the important nineteenth-century so-called Hindu reformers, is relevant here. On the basis of his religious teachings, a successful welfare organization was established, offering services that include literacy projects, agricultural extension services and disaster relief. Vivekananda was a disciple of the nineteenth-century Bengali saint Sri Ramakrishna and was one of the first Hindu teachers to travel to the West. His particular brand of Hinduism stressed a 'formless' version of the divine, which appealed to a western audience who would have been suspicious of the iconistic and polytheistic styles of Hindu belief and practice that were prevalent in India. Moreover, he was critical of the injustices of the caste system and advocated improved rights and conditions for women.

In 1897 he founded the Ramakrishna Mission (RKM), a socio-spiritual welfare organization that now has branches throughout the world. He was interested not only in attracting westerners to Hinduism but also in gaining their financial support for his 'development' work in India. For the Ramakrishna Mission, social service is a *sadhana*, a religious path, based upon the (spiritual) realization that every person is divine, and therefore to serve others is to serve God (Miller 1999: 124; Plates 4.4 and 4.5). As one Indian female development worker explained:

> Persons who join RKM join neither for money, nor fame, nor for anything else. Their motive is essentially to serve. Swami Vivekananda had preached 'one

cannot get salvation if one does not try to seek the salvation of his brothers'. ... In looking at the development activities of the World Bank, or of UN agencies, or of government projects, projects especially meant for the poor, these have often failed for there is neither a sense of 'service' nor 'renunciation', and often a lack of understanding of the essence and soul of the poor people in a given culture. Swami Vivekananda and Mahatma Gandhi had developed their understanding of the poor by living amongst the poor, and listening and interacting with them.[7]

Is Hinduism an obstacle to economic development?

In the post-independence period there were some attempts to explain India's lack of economic development as a product of Hinduism. According to this discourse, not only had the influence of Hinduism meant an elevation of the spiritual over the material, which had stunted India's economic growth, but Hinduism had also given rise to a culture that was obedient to authority and rules (e.g. the rules associated with the behaviour of different castes). This, it was argued, had tended to make 'Indians' less inclined to innovation and risk-taking, stifling the creative impulses necessary to compete in a modern global economy. It was argued that this 'Brahmanical culture' or the 'Brahmanical

Plate 4.4 *Children's Health Service Programme, Ramakrishna Mission*

Source: Ramakrishna Mission, Delhi.

The Children's Health Service Programme is one of a number of projects that the Ramakrishna Mission is undertaking between 2011 and 2014 in memory of the 150th anniversary of the birth of Swami Vivekananda.

Plate 4.5 *Medical examination of children as part of the Children's Health Service Programme, Ramakrishna Mission*

mind' – where Brahmin refers to the priestly caste – is the antithesis of modern development and progress (Moddie 1968; Dasgupta 1977).

This body of literature can be positioned within the Weberian tradition. Max Weber posited the theory of the 'Protestant work ethic' to explain the rise of capitalism in Northern Europe and he also explained the 'slower' economic development in India in terms of its cultural and religious values (1930 [1904–1905]). Debates about links between Hinduism and India's economic development became particularly popular from the 1950s onwards. These questions were pressing at the time, since India had just emerged from colonialism and all eyes were on the first post-independence government, which, in the face of mass poverty, was adopting a socialist model for development. Development concerns, combined with orientalist attitudes towards Indian religiosity and culture that were prominent during the colonial period, produced what now seems to be a rather essentialist and biased interpretation of the role that religion has played in India's economic development. The North American anthropologist Milton Singer (1956) has also written on this topic, and while he agrees that there is something more renunciatory and less materialistic (more 'spiritual') about Indian culture, he critiques any attempt to draw a hard-and-fast distinction between 'spiritualism' and 'materialism'. He suggests that the implication that materialism has no value in India or that spiritualism and materialism are mutually exclusive is perhaps more commonly drawn by westerners than by Indians (ibid.: 82).

Nonetheless, although India has modernized, not everyone has benefited equally, and the extent to which inequality may, at least partially, be explained by the persistence of certain religio-cultural forms in society (e.g. the caste system) continues to be of interest.

Hindu economics: the Gandhian influence

In contrast to the idea that Hinduism impedes India's development, Box 4.11 illustrates an alternative view: that a distinct and progressive approach to economics can find support within Hinduism.

The literature on the topic of Hindu economics focuses on key figures such as Sri Aurobindo (1872–1950), Rabindranath Tagore (1861–1941) and Mahatma Gandhi (1869–1948), noting that they all emphasized the ways in which material dimensions of life are secondary and inferior to the spiritual. However, for them this is viewed not as problematic (as it is in the 'Brahmanical mind' discourse) but as offering an alternative that is considered to be more appropriate for an Indian setting. Sri Aurobindo, Tagore and Gandhi were all prominent

---

Box 4.11

Hindu economics

In 1998 the popular publication *Hinduism Today* published an article called 'Ethical Economics' which argued that '[a]t the heart of Hindu economics is this: the test of every policy is not profit, employment or growth, but how it strengthens family and community, individual character and sensitivity' (Pandaya 1998). It cites the Hindu trade union leader and activist Dattopant Thengadi (1920–2004), who in his book the *Third Way* argues that:

> Hindu economics is fundamentally different from Western economics. To start with, he says, in the West economics is treated as a separate discipline. But within Hinduism, economics falls under *artha*, one of the four legitimate aims of life: *dharma* (righteousness), *artha* (wealth), *kama* (pleasure) and *moksha* (liberation). In Western economics, human beings are too often regarded as essentially economic beings, carrying out economic activities, producing goods and, in turn, consuming goods to complete the economic cycle. . . .
>
> In contrast to the mechanical approach of Western economics, Hindu philosophy holds that human beings are not just physical entities to be kept happy by producing and consuming. Rather, humans comprise physical, mental and spiritual aspects, and for the happiness of an individual all three should be taken into consideration. When you apply these fundamental beliefs, the kind of economics you get is very different.
>
> (Pandaya 1998)

in the *swadeshi* movement of the early twentieth century, the 'free India' movement that sought to end British colonial rule. Their combination of Hindu spirituality with lifestyle models drew upon what each saw as crucial for the future of an independent India. Each sought to establish a Hindu economics that differed from capitalism, where in the latter case it is profit, rather than spiritual development, that drives economic activity. They are also well known for their association with ashrams – communities that are based upon a combination of spiritual teachings (rooted in Hinduism) and attempts to live in community-oriented and self-sufficient ways (Minor 1998). Tagore, for example, set up the Visva Bharati University in Santiniketan, West Bengal, in the same place where his father (the spiritual leader Debandranath Tagore) had established an ashram in 1863 (Dutta and Robinson 1997).

While there is a substantial literature on all these figures, and their impact has been felt across India and beyond, it is the writings and activities of Gandhi that have been most influential. Gandhi professed a lack of trust in industrialization and capitalist expansion, promoting instead an economic system based upon the traditional Indian village system, with an emphasis on local production, crafts and self-sufficiency. The image that he adopted for an independent India (and which still appears on the Indian flag) was that of a spinning wheel (Plate 4.6) On the one hand, this was a reaction against the British export of cotton

Plate 4.6 *Gandhi spinning, late 1920s*

Box 4.12

Ralegan Siddhi and Anna Hazare

Since 1975 the village of Ralegan Siddhi, near Pune, in the state of Maharashtra, has been led by a man called Anna Hazare. Inspired by the teachings of Swami Vivekananda and Gandhi, he transformed his village from a place ravaged by drought and poverty into a prosperous village based upon 'Hindu economics, governed not by profit motives but by dharma, encompassing not just the value of goods, but also of lives' (Pandaya 1998). Using traditional methods of water harvesting and soil conservation, he enabled people to transform a drought-prone area: literacy rates rose, alcohol consumption fell and people contributed excess grain to a 'grain bank' for communal use when grain is scarce. The village temple was reconstructed, serving as a focus for both village meetings and worship.

---

from India to the mills of northern England, while on the other it symbolized a vision of a self-sufficient India (Pani 2002; Narayan 1970; Sharma 1997).

The legacy of this thought, which was rooted in the swadeshi movement and Hindu spirituality, lives on in the present day. However, unlike the promotion of Islamic economics as part of the twentieth century Islamic revival, which has come to shape much economic activity in the Muslim world, Hindu economics has been less far reaching. Hindu reformers from a similar period did not systematically pursue a focus on economic development (perhaps reflecting the anti-materialist stance in much Hindu thinking) and the Independence government of India was little swayed by religious approaches to economic matters, instead adopting a secular socialist model of development. In 1991 the economy was liberalized and capitalist models of development introduced. Nonetheless, there are some examples where individuals attempt to live by principles that reflect the ideas of these earlier Hindu thinkers, including the continuation of the ashrams and communities inspired by their thought. Box 4.12 provides an example of one such modern community – Ralegan Siddhi – which is based upon Hindu economics and inspired by the teachings of Swami Vivekananda and Gandhi.

Christianity

The Bible (comprising what Christians call the Old and New Testaments) is the main source of authority for Christians. The New Testament, which is concerned with the life and teachings of Jesus, the founder of Christianity, is more important than the 'Old Testament', which is central to Judaism in the form of the Torah (the Qur'an also contains material found in the Bible). Unlike Hinduism, Christianity and Islam are essentially missionary religions that have gained followers throughout their histories as they have spread to different parts of the globe. By contrast, Hindus historically lived mainly in India (also some in Java and Bali), until migration to East Africa, the Caribbean, Europe and North

America in the nineteenth and twentieth centuries, and today the vast majority of Hindus still live in India. While the aim of Christian missionaries was to convert people to Christianity, and colonialism opened up new opportunities for conversion, missionaries also undertook various sorts of social activities that we would now call 'development', including mass education; health care; improving women's situation; anti-caste, anti-slavery and anti-alcohol campaigns; economic and livelihood initiatives; and care for the poor. While these activities reflected Christian teachings about compassion, for instance, they also served the purpose of rendering local people more favourable to the idea of conversion. This remains an issue to today, where many in the development community are critical of Christian Evangelical organizations that combine preaching and conversion with development (see Box 8.2 on p. 209 for an examination of local and international Christian FBOs and charities; see Box 9.1 on p. 235 for a discussion of Samaritan's Purse, a controversial Evangelical Christian FBO).

There are a range of Christian denominations, often divided into three main branches – Catholic, Protestant and Orthodox – Catholicism being the largest, with over 1.1 billion members (Catholic Herald 2011) out of a total worldwide Christian population of around 2.2 billion (Foreign Policy 2007). Protestantism itself comprises very many denominations that formed following the Reformation in the sixteenth century, as a number of Churches across Europe split from the Catholic Church. The Orthodox denominations are far older, and in common with Catholicism consider that they represent the original Christianity that can be traced back to Jesus' apostles. As with Hinduism and Islam, there have been attempts to unify different denominations. A world missionary conference held in 1910 in Edinburgh gave rise to the Ecumenical Movement, which resulted in the formation of the World Council of Churches (WCC) in 1948. The Ecumenical Movement is centred on but also extends beyond the WCC – which has 350 Churches in its membership, excluding the Catholic Church, which, although not a member, does work closely with the Council.

Attitudes towards wealth

Christian teachings on development typically start with the example of Jesus. As Kim writes, he

> declined earthly wealth and status to serve others (e.g. Mark 10:45; Jn 13:1–20) . . . [and]challenged those who exploited the poor (e.g. Mt 23:23; Mk 11:15–19). He upheld religious laws that advocated justice for the oppressed, and questioned or reinterpreted others that did not (e.g. Mt 5:21–48), held up as an example those who used their resources to help the needy across religious and social barriers (e.g. the Good Samaritan, Lk 10:25–37) and, in a central act narrated in all four gospels, he shared out food to the crowd (Mt 14:13–21; Mk 6:30–44; Lk 9:12–17; Jn 6: 1–14).
>
> (2007: 5–6)

Jesus and his followers lived by the moral teachings of the Old Testament, which were centred on the 'Ten Commandments', which forbid greed and corruption and the unbridled pursuit of wealth. Through his death and subsequent resurrection he is believed by Christians to have atoned for human sin, and the Church that formed after Jesus' death established itself to live out his teachings and to await for the return of Jesus, when humans will 'live harmoniously with one another and with the created world (Rev 21)' (ibid.: 7).

Christians have differed in their understanding of the relationship between the spiritual and the material, with some adopting an other-worldly approach whereby they avoid attachment to material things in order to become closer to God (e.g. monks and nuns). Others have seen material success as a sign of being chosen by God, a feature that was seized upon by Weber as a crucial link in his theory of the 'Protestant work ethic', which he believed contributed to the rise of capitalism. Kim draws attention to a distinction between so-called social Christianity, a liberal Protestant tradition emerging in the nineteenth century that is progressive and modernist, focusing on 'Jesus' ethical teaching to love God and neighbour rather than on doctrine' (2007: 14), and Protestant-based 'Evangelical Christianity', which began to emerge in the seventeenth century and required the need for converts to be 'born again' through accepting Jesus as their saviour. Whereas 'social Christianity' (which has found expression in different forms, for instance as 'Christian socialism' in the United Kingdom and the 'social gospel' in North America) is more likely to underpin movements that aim to emulate Jesus' example in helping the poor and marginalized, Evangelical Christianity is more comfortable with capitalism, even openly preaching a 'prosperity gospel', a message that success in material terms is a sign of God's favour.

Another style of Christianity that is also Evangelical and preaches a 'prosperity theology' that encourages families and individuals to create wealth is Pentecostalism, dating from the early twentieth century (Martin 2002: 14–16; Cox 1996: 213–241). Pentecostal churches seek to encourage attitudes towards wealth that are consonant with capitalist models of economic development. Pentecostals are also likely to have a doctrine of 'tithing', where 10 per cent of family income is given to the Church. Kim points out that 'the largest churches (such as the Yoido Full Gospel Church, South Korea) are now in a position to found education establishments and produce development programmes of their own' (2007: 29). While Pentecostalism shares much in common with Evangelical theology, it differs in its style of worship, which involves 'awareness of the immediate presence of God by the Holy Spirit, expectation of some sign of miraculous/powerful intervention, and spontaneous congregational participation (Anderson, 2004, pp.1–15)' (Kim 2007: 28). Pentecostalism is thought to be the world's fastest-growing religion, by conversion, and in many parts of the globe has come to underpin the 'churches of the poor'. In Africa they are also sometimes known as '"African independent (indigenous or initiated) churches" (AICs)', to draw attention to their establishment as 'expressions of independence from colonial masters' (ibid.: 27). According to Kim, Pentecostal churches tend to be conservative in their values and also upwardly mobile owing to the opportunities they offer, including 'a drug and alcohol-free environment,

Table 4.1 *Estimated size of renewalist populations (in percentages of the total population)*

	Pentecostals	Charismatics	Total (renewalists)
United States	5	18	=23
Latin America			
Brazil	15	34	=49
Chile	9	21	=30
Guatemala	20	40	=60
Africa			
Kenya	33	23	=56
Nigeria	18	8	=26
South Africa	10	24	=34
Asia			
India (localities)	1	4	=5
Philippines	4	40	=44
South Korea	2	9	=11

Source: Pew Forum (2006).

high standards of personal morality that aid stable family life, opportunities and encouragement in education and leadership, and a network of support' (ibid.: 28). The umbrella term 'renewalist' is used in Table 4.1 (which provides figures on the estimated size of renewalist populations in different countries) to refer to Pentecostals and those within other Christian denominations who are 'charismatics' (i.e. have experienced the 'in-filling' of the Holy Spirit, yet are not Pentecostals; Plate 4.7).

Attitudes towards poverty

The primary biblical understanding of poverty is as destitution, drawing attention to the material and social aspects of 'poverty'. However, as Rylaarsdam demonstrates, as the biblical tradition develops, 'poverty' and 'the poor' 'began to serve as theological symbols that hinted at the human position and attitude that should govern the relationship between God and man [*sic*] . . . illustrated by the first of the Beatitudes, "Blessed are the poor in spirit for theirs is the kingdom of heaven" (Mt. 5:3)' (1968: 13). Thus, voluntary poverty, in the sense of emulating the recurring theme of renunciation in Jesus' teaching, is regarded by some Christians as a virtue and as necessary for reaching God. As Jesus tells us in Mark 10.25, a rich man trying to enter the Kingdom of Heaven is like a camel trying to pass through the eye of a needle. While the relevance of this teaching in practice for those who are already poor is questionable, we do also find understandings of involuntary poverty as God's will to serve the purpose of discipline, testing or punishment (Job 1). Moreover, Mark 14.7 tells us that 'the poor are always with us', reflecting poverty as part of 'a necessarily unequal world' (Kim 2007: 31). Although, as with other religions, we find such interpretations of poverty verging on the fatalistic, these are accompanied by the view

Plate 4.7 *Pentecostalism in Uganda*

Source: AFP/Getty Images.

Ugandans attending a service led by Pastor Robert Kayanja (one of Uganda's wealthiest and most controversial Evangelical pastors, having been at the centre of a sexual abuse accusation in 2009) in Mbarara, western Uganda's largest town, on 23 August 2008.

that it is something to be eradicated and is at root the product of 'the corruption of human beings or society by sin, or the outcome of political and economic injustice and oppression (cf. John 9:1–3)' (Kim 2007: 31; see Taylor 2003).

Helping the poor: poverty alleviation

An emphasis upon Jesus' social teachings upon which Christians should model their behaviour is found across all Christian denominations. Individual Christians have a moral responsibility to follow Jesus' example, in helping the poor and reducing their suffering. This is also something taken seriously by Christian churches as institutions. As Kim writes, within the Protestant tradition, in the late nineteenth century '[s]ocial Christianity arose specifically from realisation of the needs of the urban poor of industrial society. The movement had different strands such as the "social gospel" (in North America), "Christian socialism" (in Britain), or "religious socialism" (on the Continent)' (2007: 15). Moreover, in contrast to Islam and Hinduism, Christianity (as the predominant religion in the West) became directly drawn into the newly emerging donor-driven development project following the Second World War. As Kim points out, the World Council of Churches (WCC) encouraged members to 'participate in the post-War re-construction, nation-building and development agenda, and cooperate with governmental and non-governmental bodies for these specific purposes' (2007: 18). Taylor (1995) demonstrates how the WCC was concerned to help the poor as central to its work and in its early days 'accepted the model of transnational economic development by transfer of technology, as the means to achieve this' (Kim 2007: 19). However, as with the western development model more broadly, by the 1960s and 1970s, under the influence of Marxism and churches based in the South, this approach was called into question for being elitist and serving western interests (Taylor 1995: 47–75). At a conference in Bangkok in 1973 the WCC adopted a '"comprehensive" understanding of salvation (or development) as liberation in four inter-related dimensions: economic justice, human dignity, human solidarity, and "hope in personal life" (World Council of Churches 1973)' (Kim 2007: 19). While traditionalists attacked this for lacking an explicitly Christian message, the WCC began to support movements concerned with challenging injustice globally and in the 1980s also adopted the 'option for the poor' of liberation theology (see pp. 116–118). In 1983 it added 'the integrity of creation' to the agenda of a ten-year programme on 'justice and peace'.

Within the Catholic Church, social teaching has been developed and system-atized over the Church's two-thousand-year history. However, since the Second Vatican Council (1962–1965), which addressed and reinvented the relationship between the Church and the modern world, more recent social teaching reflects contemporary development discourse (for example, see the website of CAFOD for research papers applying Catholic teaching to development issues; Deneulin and Bano 2009: 136–142). In his 1967 encyclical *Populorum Progressio* ('On the development of peoples'), Pope Paul VI embraced the term 'development' as 'the new name for peace'. The encyclical draws attention to the pros and cons of colonialism, the widening gap between rich and poor globally, the prevalence of social unrest born of deprivation and the role of the Church in bringing about 'authentic development'. As Box 4.13 illustrates, 'development' here is not restricted to 'economic growth' and instead must involve the development of the whole person.

Moreover, under the influence of liberation theology (see pp. 116–118), since the papacy of John Paul II (1978–2005) the Church has expressed 'a special solidarity with the poor', implying 'resistance to exploitation and the obligation to release the poor from debt' (Kim 2007: 12). As Box 4.14 shows, in *Sollicitudo Rei Socialis* ('On social concern', 1987) John Paul II outlines this in the light of the recognition that 'today more than in the past, the Church's social doctrine must be open to an international outlook' (1987). Nonetheless, Catholic teach-ings are particularly weak in the area of gender equality; they rarely mention women, and when they do, it is in terms of their biologically predetermined roles. Moreover, the use of masculine words to refer to the whole of humanity is striking in these Catholic documents.

Box 4.13

Populorum Progressio (1967)

In his 1967 encyclical Pope Paul VI stated that:

1. The progressive development of peoples is an object of deep interest and concern to the Church. This is particularly true in the case of those peoples who are trying to escape the ravages of hunger, poverty, endemic disease and ignorance; of those who are seeking a larger share in the benefits of civilization and a more active improvement of their human qualities; of those who are consciously striving for fuller growth. . . .

14. The development We speak of here cannot be restricted to economic growth alone. To be authentic, it must be well rounded; it must foster the development of each man [sic] and of the whole man [sic]. . . .

20. . . . This is what will guarantee man's [sic] authentic development – his transition from less than human conditions to truly human ones.

Source: www.vatican.va/holy_father/paul_vi/encyclicals/documents/hf_p-vi_enc_26031967_populorum_en.html (last accessed 8 September 2012).

Box 4.14

Sollicitudo Rei Socialis (1987)

In his 1987 encyclical Pope John Paul II stated that:

42. Today more than in the past, the Church's social doctrine must be open to an international outlook . . . the option or love of preference for the poor. This is an option, or a special form of primacy in the exercise of Christian charity, to which the whole tradition of the Church bears witness. It affects the life of each Christian inasmuch as he or she seeks to imitate the life of Christ, but it applies equally to our social responsibilities and hence to our manner of living, and to the logical decisions to be made concerning the ownership and use of goods. . . .

46. . . . Development which is merely economic is incapable of setting man [sic] free, on the contrary, it will end by enslaving him [sic] further. Development that does not include the cultural, transcendent and religious dimensions of man [sic] and society, to the extent that it does not recognize the existence of such dimensions and does not endeavor to direct its goals and priorities toward the same, is even less conducive to authentic liberation. Human beings are totally free only when they are completely themselves, in the fullness of their rights and duties. The same can be said about society as a whole. . . .

Source: www.vatican.va/holy_father/john_paul_ii/encyclicals/documents/hf_jp-ii_enc_30121987_sollicitudo-rei-socialis_en.html (last accessed 14 December 2011).

Liberation theology

Liberation theology has its roots in 1960s Latin America, when Catholic priests encouraged the poor to read the Bible together in 'base communities' in the light of their social and economic condition (Plate 4.8). Influenced by Marxism and the pedagogical theories of Paulo Freire, they saw teachings in the Bible that spoke directly to their situation (Hennelly 1990). Liberation theology was critical of the Roman Catholic hierarchy for supporting various military regimes in Latin America as well as the inherent injustice of global capitalism, and gave people 'a biblical basis for Christian action on behalf of the poor' (Kim 2007: 20; Plate 4.8). It emerged as part of the Latin American response to the Second Vatican Council, after which local bishops were granted greater autonomy. At the first conference of Latin American bishops held after Vatican II, in 1968 in Medellín, Colombia, a Peruvian Dominican priest called Gustavo Gutiérrez gave a speech that he later published as *A Theology of Liberation* (1973 [1971]), and liberation theology was born (Deneulin and Bano 2009: 143). One of the key features of this new theology was what was called the 'preferential option for the poor': the individual Christian should adhere to God's commandment to help the poor, oppressed and marginalized.

The phrase 'preferential option for the poor' was first used by Father Pedro Arrupe, Superior General of the Society of Jesus (Jesuits), in 1968 in a letter to

Plate 4.8 *Liberation theology in El Salvador*

Source: Alison McKellar from Camden, Maine (Creative Commons). Online, available at: www.flickr.com/photos/alison mckellar/2105760285/.

This mural depicts Archbishop Monseñor Óscar Romero, perhaps the best-known religious figure and human rights activist in El Salvador. It represents the important role the Catholic Church played in the struggle for peace. Romero, although conservatively inclined, became one of the most high-profile Catholic figures associated with liberation theology. He was shot and assassinated on 24 March 1980 during a mass at a cancer hospital in San Salvador.

the Jesuits of Latin America, but has become influential across different Christian denominations and beyond. While the Catholic Church condemned the use of Marxist social theory and socialist solutions proffered by liberation theologians, it did adopt the idea of the 'option for the poor' into its social teaching (Hennelly 1990). However, in the 1980s the Vatican denounced liberation theology on a number of counts, including the emphasis of some liberation theologians on Marxism, and hence atheism, and also the seeming support shown by some for guerrilla forces. According to Deneulin and Bano, the Church removed 'all Latin American bishops who were openly supportive of liberation theology and replaced them with clerics who did not view political and social engagement against poverty and injustice as being the church's role' (2009: 145). Coupled with the fact that the transition to democratic rule meant that an oppositional Church became obsolete, the rise of Pentecostalism, which led the Church to adopt the same 'charismatic, devotional and apolitical' stance in order to compete for adherents, has led to the demise of liberation theology (ibid.: 145).

Nonetheless, liberation theology has been widespread and influential, even if the original movement now lacks a strong base. Because the movement was justified by reference to the Bible, and not Catholicism per se, it was able to 'spread through the Ecumenical movement to Protestants, and influenced Evangelical churches too' (Kim 2007: 20). The adoption of the 'preferential option of this poor', according to Kim, 'suggested a new pattern of action combined with theological reflection: identifying with the poor, listening to them, re-reading the Bible in the light of their experience, and joining with people's movements to bring about change from the grassroots' (ibid.: 20). As an approach for uplifting the downtrodden, oppressed by religious, political and social hierarchies, it has influenced emerging forms of liberationist feminist theology. In Asia it underwent further development and has been applied not only to economic but also to social and cultural discrimination. In India, for instance, liberation theology has become one of the influences on Dalit movements (Stanislaus 1999). The influence can also be seen in the widespread support of the different churches in Britain for the Jubilee 2000 Debt Campaign (Chapter 2) and the subsequent Make Poverty History movement. Kim suggests that it has enabled 'older liberal Protestant agencies such as Christian Aid . . . to gain the support of newer Evangelical bodies such as TEAR Fund, and Roman Catholic organisations such as CAFOD (as well as some other faith groups), to speak to governments with one voice against poverty' (2007: 21).

Economic development

As was discussed in Chapter 2, mainstream economic theory has tended to pay very little attention to religion. However, in contrast to Islam and Hinduism, Christianity has played a closer role in shaping western economic theory and practice, due to its status as the primary religion in Europe. Oslington (2008) demonstrates the close ties between Christian theology and the emergence of modern political economy in the eighteenth and nineteenth centuries in the economic theories of key figures such as Thomas Malthus, Adam Smith,

William Paley and David Ricardo. He writes that although 'natural theology legitimated political economy, shaped economic theory and provided a common language for economists and theologians through this period' (2008: 5), during the nineteenth century 'forces of specialization and professionalization operated to separate all the sciences from theology' (2008: 8).

Despite the fact that by the nineteenth century, theology had been marginalized in mainstream economic thought, the churches and their theologians have continued to ask questions about economic issues. Moreover, as Oslington suggests, since the mid-1970s there has been more discussion of the relationship between theology and economics at a time when an 'interest in Christian critiques of economics coincided both with a larger crisis of confidence in economics and with a revival of interest among Evangelicals in social issues' (2008: 15). This also includes increased Catholic comment on the role of the market, for example in the work of the North American Catholic philosopher Michael Novak, and also Pope John Paul II's *Centesimus Annus* (1991), in which he called for a more holistic understanding of the person, one that goes beyond economics or class (see Box 4.15). The influence of the so-called reformed tradition within Protestant Christianity in the United States, with work from Donald Hay and Kim Hawtrey, has also been significant (Poole 2010; Harper and Gregg 2010; Sung 2007).

In the modern era, one of the main tensions that exists between different Christian approaches towards economic development is support for capitalism, on the one hand, or support for socialist models, on the other. In the Catholic Church, however, modern Catholic social teaching (which dates from an 1891 encyclical of Pope Leo XIII *Rerum Novarum*, 'The condition of labour') explicitly steers a path between socialism or Marxism and capitalism (Kim 2007: 12). Whereas the former does not recognize the sacredness of the human person, the latter neglects the social aspects of the human person. The Catholic Church has not adopted a new economic system, nor does it reject capitalism as a whole,

---

Box 4.15

Centesimus Annus (1991)

In his 1991 encyclical Pope John Paul II stated that:

> To [the inefficiency of the economic system] must be added the cultural and national dimension: it is not possible to understand man [*sic*] on the basis of economics alone, nor to define him simply on the basis of class membership. Man [*sic*] is understood in a more complete way when he is situated within the sphere of culture through his language, history, and the position he takes towards the fundamental events of life, such as birth, love, work and death. At the heart of every culture lies the attitude man [*sic*] takes to the greatest mystery: the mystery of God. Different cultures are basically different ways of facing the question of the meaning of personal existence.

Source: www.vatican.va/holy_father/john_paul_ii/encyclicals/documents/hf_jp-ii_enc_01051991_centesimus-annus_en.html (last accessed 8 September 2012).

but 'tends to advocate a basic minimum for everyone coupled with opportunity for individual accumulation of wealth as part of human development' (ibid.: 12).

While the Catholic Church has tended therefore to adopt a middle ground, as we have already seen, the 'prosperity gospel', found within some versions of Evangelical Christianity and Pentecostalism, preaches that God created the world's riches for people to enjoy, that he rewards believers with riches and that the individual should make tithes or offerings to the churches. Hackett, for instance, describes how in West Africa an explicit link is made between membership of Pentecostal churches and liberation from poverty, and argues that there is a 'close affinity between the type of religious multinationalism, religious enterprise, values and competitive pluralism engendered by the charismatic revival in Africa and global, particularly Western, capitalist forces' (1995: 199; see also Pew Forum 2006; Martin 2002). While the churches maintain that the money people give is for the purposes of mission as well as charity, there is often a lack of transparency in how funds are used, and some pastors have been found guilty of using money for their own ends. Critics attack this theology for not reflecting how Jesus actually lived, as well as for being exploitative of the poor.

At the other end of the spectrum we find support for approaches to economic development that combine socialism with Christianity. An important element of the broader 'social Christianity' movement that had emerged by the late nineteenth century was 'Christian socialism', a term that was first used by a group of men in Britain around 1848 following the failure of the Chartist petition to Parliament, including Frederick Denison Maurice, Charles Kingsley and John Ludlow (White and Hopkins 1976: 26). Chartism was a working-class movement for political reform, including the right for all men to vote. However, as White and Hopkins write, at that time these individuals understood socialism to mean 'little more that the principle and practice of cooperation as opposed to economic competition . . . socialism itself had not developed into that more precise economic thought which is understood by the world today' (ibid.: 26). For these early Christian socialists, 'socialism' was the true meaning of Jesus' teaching as a response to the poverty and oppression that the Industrial Revolution had brought to the working classes. The Christian socialist tradition remained influential in Britain until the 1980s, but the extent to which the Marxist version of socialism was adopted by these social Christians varied: some were committed to a style of Protestant Marxism that advocated the overthrow of capitalism and the prohibition of private property, whereas others considered socialism and social Christianity to imply cooperation and greater sharing of resources, albeit within a reformed capitalist economy.

Summing up

The aim of this chapter was to better understand how we might identify religious approaches to 'development' and the significance of these for mainstream development policy and practice. More specifically, it sought to address the following questions: *How might members of religious traditions approach 'development'? In what ways does religion influence visions of development as well as the policies and strategies, to reach those visions? And are these religious approaches significantly different from secular approaches?* In the first section of the chapter we examined the extent to which it is possible to identify religious approaches to development, considering the heterogeneity of religious traditions, and how we might go about this. In particular, the importance of carrying out ethnographic research to identify 'lived' or 'popular' religion was emphasized. In the second section, we then examined the purpose and desirability of economic development, from the point of view of commentators within Islam, Hinduism and Christianity, key dimensions of which are the pursuit (and value) of wealth and how to tackle poverty, including whether/how the redistribution of wealth should occur via individual efforts and/or the creation of alternative more equitable economic systems.

In contrast to western economic approaches to development, the three religions surveyed (Islam, Hinduism and Christianity) have strong traditions of renunciation or moderation in the human pursuit of wealth. So-called world renunciation is a practice found across religious traditions, where the individual chooses to reject the comforts of the material world, as they are seen as coming between the individual and the highest goals of the religion concerned. In Islam we find the Sufi mystics, focusing on 'the internal or more spiritual aspects of Islam . . . the seeking of closeness to God through particular rites, such as the invocation of saints or particular dance practices. . . . Asceticism and subduing one's own ego are also key practices' (Kroessin 2008: 25). In Hinduism we find the wandering *sadhus* (holy men), having no possessions or fixed abode. And in Christianity there are traditions of monks and nuns, living away from the distractions of modern societies.

While these renunciate traditions have an important place in each religion, such a focus on non-materialism is hardly practical for the survival of a religion, and lay practitioners tend to involve themselves more fully with the world. Nonetheless, these ideas about material things standing between the individual and religious truth permeate even at the level of lay religion, and religious traditions often aim to encourage the individual to live in the world while at the same time not placing ultimate value on the pursuit of wealth at the expense of other aspects of life. Thus, members of each tradition surveyed often claim that the approach to development from within their religion differs from mainstream approaches in the Global North, particularly where the focus upon economic development at the expense of other ways of viewing development, including 'spiritual' development as well as broader notions of well-being beyond financial prosperity. Of course, there are exceptions to this tendency towards moderation of economic wealth, most notably in the Christian 'prosperity gospel', popular particularly within Pentecostalism.

While religion has been marginalized from mainstream economic theory, religious traditions have continued to develop their own 'economic theologies'. There are different views about the extent to which capitalism needs to be reformed, but overall, many of the religious perspectives examined in this chapter exhibit an awareness of the inequalities inherent within the capitalistic system and seek to address these. In terms of assessing the implications of this discussion for mainstream development policy and practice, one insight that can be gleaned from this chapter is the limitations of approaches to development that focus upon economic growth as an end in itself. While shifts towards human development and the capabilities approach since the 1980s have begun to move development theory and practice in this direction, donor-driven approaches to development continue to rely upon capitalist models of economic growth. Moreover, it is important for international development actors to have knowledge about other cultural attitudes towards development-related issues such as wealth and poverty in order to ensure that the projects and programmes they initiate are the ones that are most likely to help people improve their lives. For instance, understanding religious attitudes towards poverty, in terms of its definition, causes and significance, can be crucial to gaining an overall understanding of the ways in which particular communities view the poor and the extent to which they are likely to have mechanisms for supporting them. This knowledge could also be the basis for developing partnerships with religious charities and institutions.

However, as was noted at the end of the previous chapter, the engagement with religions by international development actors is more likely to be promoted when religious traditions share their 'understanding of what constitutes rationality, progress, social justice and modern economic development' (Thomas 2005: 220). One of the challenges or barriers facing development actors in taking religion seriously and in thinking about ways in which development theory and practice can more closely connect with the 'moral base' of society (Wilber and Jameson 1980) is how to negotiate and deal with multiple and often conflicting religious voices, where religions conflict not only within and between themselves but sometimes also with the main goals of the modern development project. However, if these development actors cling to a secular and 'universalist' vision, then are they not guilty of ethnocentric bias and the neglect of alternative worldviews? Neither, though, does 'cultural relativism' (i.e. the view that all cultures must be treated on their own terms and it is wrong for outsiders to interfere) offer much in the way of solution, since there is much within different cultural and religious traditions that can legitimate oppression. In the following chapter we will explore this tension between universalism and cultural relativism with respect to the relationship between religions and human rights, and will examine the implications of this for 'human rights-based approaches' (HRBAs) to development. While HRBAs have tended to avoid engaging with religion in pursuit of rights, since there is often an assumption that religion is more likely to act against rights, we will explore the argument that approaching human rights from within religious frameworks could help bridge the chasm between the theory and realization of rights in development.

Questions for discussion

1 What challenges are faced in attempting to identify 'religious ap   ....s to development'?

2 With respect to each of the religious traditions discussed in this chapter, what perspectives on the following can be identified?

   i the pursuit of wealth;
   ii the nature and causes of poverty;
   iii the motivations and mechanisms for alleviating poverty and redistributing wealth;
   iv the purpose and desirability of economic development.

3 In your opinion, in what ways can we demonstrate that these approaches to wealth, poverty and economic development differ from mainstream secular approaches to development?

Recommended reading

Clarke, Matthew (2011) *Development and Religion: Theology and Practice*, **Cheltenham, UK and Northampton, MA: Edward Elgar Publishing Limited.** This book explores how the world's five major religions – Hinduism, Buddhism, Judaism, Christianity and Islam – understand and practice 'development' through an examination of their sacred texts, social teaching and basic beliefs.

The following texts are introductions to the 'world religions' discussed in this chapter:

Armstrong, Karen (2001) *Islam: A Short History*, **Sheffield: Phoenix.**

Knott, Kim (2000) *Hinduism: A Very Short Introduction*, **Oxford: Oxford Paperbacks.**

McGrath, Alister E. (2006) *Christianity: An Introduction*, **2nd edn, Malden, MA: Blackwell.**

Rodrigues, Hillary (2006) *Introducing Hinduism*, **New York: Routledge.**

Ruthven, Malise (2000) *Islam: A Very Short Introduction*, **Oxford: Oxford Paperbacks.**

Woodhead, Linda (2004) *Christianity: A Very Short Introduction*, **Oxford: Oxford Paperbacks.**

Websites

http://hindungos.blogspot.com/2009/09/hindu-ngos-worldwide.html Details are available on this website about a book that lists Hindu NGOs worldwide.

www.cafod.org.uk CAFOD is the official Catholic aid agency for England and Wales. This website provides resources for different users, including policy papers that apply Catholic teaching to development issues.

www.islamic-relief.org.uk Islamic Relief is a UK-based aid and development charity. The website contains details about the work of the organization as well as publications linking Islam to development concerns.

⬤5 Human rights, religions and development

This chapter will:

- ◉ address the questions: *Are religious teachings compatible with human rights? What are the implications for 'development'?;*
- ◉ examine the ways in which considerations of human rights have entered mainstream development discourse, including policy debates and the design and delivery of interventions;
- ◉ assess the extent to which considerations of religion and culture have been marginalized from mainstream human rights debates and practice, including those within donor-driven development;
- ◉ discuss the application of religious teachings to human rights;
- ◉ specifically examine the tension between understandings of human rights as universal and culturally relative, and look at the role of religion in this;
- ◉ examine the extent to which an approach that looks at 'overlapping consensus' in understandings of rights from different cultural perspectives can resolve this tension;
- ◉ assess the implications of this for mainstream development policy and practice.

Introduction

There are many examples across the globe where religions seem to act against the human rights of particular groups and individuals. For instance, Christianity is influential in shaping the political discourse on homosexuality in developing countries (especially in Africa), as it is in the West. However, religious traditions are not monolithic and can also be subjected to processes of reinterpretation. It is common to find the use of liberal religious discourse, as well as secular arguments, to counter conservative or fundamentalist manifestations of religion. Yet despite the fact that religious traditions can be interpreted to support liberal human rights discourses, it is still rare to find religion incorporated into mainstream human rights debates, and some argue that this makes human rights approaches seem biased towards the values of the 'secular West'.

In this chapter we will address the following questions: *Are religious teachings compatible with human rights? What are the implications for 'development'?* Clearly, there is not a straightforward answer to these questions since, as already noted, religious traditions are not monolithic and within any religion we can find a range of positions on human rights, from those that are broadly aligned with modern universal human rights discourse to those which offer quite distinct alternatives that are considered to reflect the teachings of the particular tradition. We will explore two examples of this: the Asian values debate and its Bangkok Declaration, and the Cairo Declaration on Human Rights in Islam. This points to a tension between universalism and cultural relativism within rights discourses, and an attempt will be made to address this dilemma by examining the role of an 'overlapping consensus' approach to rights.

Moreover, what do we mean by 'development' here? Reflecting the primary focus of this book, we will be interested in examining the ways in which religious traditions are relevant to the incorporation of human rights into mainstream development policy and practice. Human rights discourse increasingly influences the way in which development organizations approach their activities, driven by the recognition that the pursuit of human rights and the pursuit of development go hand in hand. It is common for development agencies to invoke human rights as guiding their policies and programmes, for example through the adoption of 'human rights-based approaches' (Piron and O'Neil 2005). Crucially, an understanding of development itself as 'an inalienable human right' was formalized by the United Nations in 1986.

However, as Wilber and Jameson (1980) have argued, unless development policies and practices reflect the 'moral base' of society, then they are unlikely to achieve their aims and may well run the risk of doing more harm than good. While the 'moral base' of western liberal societies is no longer directly shaped by its religious heritage, with those societies now relying on logic and natural law to establish what is right and wrong, in many other cultures religion remains an important influence on people's understandings of how they should behave, the rights that they deserve and the duties that they have. Peerenboom, for instance, has acknowledged that many rights advocates across the globe

> have learned that implementation is easier and more effective *when supported by local traditions*, as confirmed by the experience of women's rights groups in Malaysia and Indonesia that have made considerable progress in their daily battles by working within their religious and cultural traditions. In general, laws that are not in accord with the values of a particular society will be difficult to enforce. Thus, *if only for purely strategic or instrumental reasons, culture, traditions, and values do matter*.
>
> (2003: 57; emphasis added)

Nonetheless, despite the fact that the relationship between religions and human rights captures an important feature of the struggle for justice faced by many of the poor in developing countries, serious consideration of the ways in which religion blocks or facilitates the pursuit of human rights has tended to be overlooked in mainstream development policy and practice. Human rights as encapsulated in the Universal Declaration of Human Rights (UHDR) are themselves strategic or instrumental tools for ensuring that people's interests are met and they have a strong appeal precisely because they are applicable to all human beings regardless of who they are. However, as Peerenboom (2003) and others have argued, if rights advocates seek to ensure that everyone across the globe can claim their rights, then they cannot ignore the role that culture, religion or politics might play in achieving this aim, with respect to both supporting and blocking rights. To pursue human rights from *within* different religious and cultural frameworks may have practical and strategic benefits, as well as recognizing the importance of people's religious and cultural traditions in determining how they view their well-being and development.

The chapter begins with a discussion of the entry of human rights into mainstream development policy debates and the design and delivery of interventions. I shall suggest that the avoidance of considerations of religion and culture in 'human rights-based approaches' (HRBAs) to development has exacerbated the chasm between the theory and the realization of human rights. We will then explore two alternative versions to the UDHR: the 'Asian values debate' and its Bangkok Declaration; and the Cairo Declaration on Human Rights in Islam. A common theme running through critiques of and alternative versions to the UDHR is that it is western and secular, and that it makes no reference to religion and culture (religion only comes into secular rights declarations in terms of the right to religious freedom). Moreover, these alternative human rights versions are critical of the emphasis in the UDHR upon individual rights above duty to the family and the wider religious community.

These two examples comprise extreme rejections of mainstream western rights discourse and deviate in crucial ways from the values of equality and democracy that underpin enshrined in the UDHR. However, there are also more moderate religious perspectives which argue that there is much overlap between western human rights and different religious value systems. The chapter will conclude with an examination of the implications of this 'overlapping consensus' approach to human rights for 'human rights-based approaches' to development. Far from suggesting a watering down of human rights principles or a transgression of religious values, the search for overlap between secular rights frameworks and religious teachings in some contexts indicates scope for fruitful engagement between development actors and religious traditions in pursuit of human rights. Many consider that this approach to rights has the potential to bridge the chasm between the theory and realization of human rights for people living in different contexts shaped by diverse religious and cultural traditions.

Human rights and mainstream development policy and practice: the emergence of the 'right to development' and 'human rights-based approaches' to development

Following the atrocities of the Second World War, the United Nations was established in 1945, and in 1948 it adopted the Universal Declaration of Human Rights (UDHR), 'a road map to guarantee the rights of every individual everywhere'.[1] The UDHR was drawn up after several years of consultation and, as

Waltz has demonstrated, although there has been a perception that it was arrived at without the participation of 'small states', in fact it involved the representatives of 'some 250 delegates and advisors from fifty-six countries' (2001: 49). The declaration consists of thirty articles, which have subsequently been elaborated upon in different international treatises and regional human rights instruments, as well as national constitutions and laws. The UDHR is one part of what has come to be known as the Universal Bill of Human Rights. In addition to the UDHR, which contains the general principles or standards of human rights, the Universal Bill of Human Rights also includes two covenants, adopted in 1966: the International Covenant on Economic, Social and Cultural Rights and the International Covenant on Civil and Political Rights (OHCHR 1996).

The importance of human rights for mainstream development policy and practice has steadily increased since the United Nations made its declaration of human rights in 1948. In particular, the shift away from the dominance of economic models and the inclusion of human development, influenced by Amartya Sen's 'capabilities approach', have meant that development policy makers and practitioners are more likely to think about human rights as relevant to their work through the adoption on 'human rights-based approaches' (HRBAs). While today the idea that the pursuit of human rights and the pursuit of development are mutually dependent shapes the policies and practices of both human rights advocates and of development actors, Alston and Robinson note that:

> [e]ndeavours to promote meaningful and productive linkages between the agendas of these two communities are hardly new. Indeed, one of the main achievements of the first World Conference on Human Rights, held in Tehran in 1968, was precisely its assertion that 'the achievement of lasting progress in the implementation of human rights is dependent upon sound and effective national and international policies of economic and social development'. Almost a decade later, in 1977, the UN commission on Human Rights gave a new impetus to these efforts by proclaiming the existence of a human right to development.
>
> (2005: 1)

The idea of the 'right to development' (RTD) was also recognized in 1981 in the African Charter on Human and Peoples' Rights, and by 1986 the United Nations General Assembly, an important forum for North–South dialogue and increasingly a means by which developing countries are able exert an influence, adopted the 'Declaration on the Right to Development':

> The right to development is an inalienable human right by virtue of which every human person and all peoples are entitled to participate in, contribute to, and enjoy economic, social, cultural and political development, in which all human rights and fundamental freedoms can be fully realized (article 1).[2]

Despite its appeal as an ideal, the RTD has proved controversial. As Marks argues, donor nations in the Global North have been concerned that the RTD has been 'used rhetorically to amplify Third World demands on the industrialized world for a transfer of resources, in the form of foreign aid or debt forgiveness' (2001: 8). Moreover, he also suggests that the United States has generally opposed the 'right to development' for ideological reasons, objecting to the notion that development should be a right rather than the outcome of 'free enterprise domestically and free trade internationally' (2004: 145). Another contested aspect of the RTD is that it recognizes economic, social and cultural rights (ESC), which have often been classified as secondary compared to political and civil rights (PC) (Hamm 2001: 1006). According to Marks, opposition to the notion that economic, cultural and social rights are equal to political and civil rights and that the state has a legal duty or obligation to realize them has been voiced, for instance, by the United States, where the view has been that ESC rights are 'merely aspirational' (2004: 147).

Despite these tensions, the RTD was reaffirmed in the 1993 Vienna Declaration and Programme of Action, a human rights declaration adopted at the World Conference on Human Rights, and also in the 2001 Durban Declaration and Programme of Action and the 2003 Millennium Declarations. Nonetheless, Piron (2002) suggests that because the very question of what it means for development to be a human right, as well as the policies to realize this, are far from settled, it is useful to separate debates about the RTD generally from human rights-based approaches (HRBAs) specifically, which have increasingly influenced the ways in which mainstream development agencies carry out their work. For instance, in 1997 the UN Secretary-General called for the mainstreaming of human rights across all United Nations agencies, and in 2003 a 'common understanding' of a human rights-based approach was adopted across UN agencies. This is outlined in Box 5.1.

According to this 'common understanding', a 'human rights-based approach' to development involves the pursuit of human rights as part of development as

---

Box 5.1

A 'common understanding' of a human rights-based approach

Within all UN agencies:

1. All programmes of development co-operation, policies and technical assistance should further the realisation of human rights as laid down in the Universal Declaration of Human Rights and other international human rights instruments.
2. Human rights standards contained in, and principles derived from, the Universal Declaration of Human Rights and other international human rights instruments guide all development cooperation and programming in all sectors and in all phases of the programming process.
3. Development cooperation contributes to the development of the capacities of 'duty-bearers' to meet their obligations and/or of 'rights-holders' to claim their rights.

Source: http://hrbaportal.org/the-human-rights-based-approach-to-development-cooperation-towards-a-common-understanding-among-un-agencies (last accessed 10 September 2012).

well as the adoption of human rights principles to guide development cooperation and programming. Specifically, HRBAs should incorporate the principles of 'equality and equity, accountability, empowerment and participation . . . approaches [that] are incompatible with development policies, projects or activities that have the effect of violating rights, and that permit no "trade-offs" between development and rights'.[3] This commitment to HBRAs is shared more widely across the development sector, and 'many bilateral donors including CIDA, DANIDA, DFID, OECD-DAC, and SIDA have made [HRBAs] a priority. So too have international NGOs such as CARE and OXFAM' (UNDP 2006: 15).

Nonetheless, as the United Nations and other agencies recognize, 'there remains a chasm between theory and practice', and 'there are, of course, many reasons why this is so, including continuing gaps in knowledge and skills, and *difficulties in translating human rights norms into concrete programming guidance applicable in diverse policy contexts and national circumstances*' (emphasis added) (OHCHR 2006: iii). One of the aims of the remainder of this chapter is to explore the argument that in order to address the chasm between theory and practice in realizing human rights as part of mainstream development, it is important to consider the role of religion and culture. However, what would doing so mean for development policy and practice, considering that religious traditions are often at odds with human rights? Would it mean compromise and a watering down of human rights principles?

Religions and human rights

Debates about the relationship between religions and human rights within the discourse of secular liberalism often consider religions as potentially problematic for the pursuit of human rights. Supporters of universal human rights argue that the very strength of contemporary human rights discourse as enshrined in the UDHR is the fact that human rights belong to every individual by virtue of their being human and regardless of their country of origin, sex, race or religion. Human rights exist objectively and are 'inalienable' – that is to say, they cannot be taken away from the individual without threatening the very essence of what it is to exist as a person. This lack of dependence on particularities provides one explanation for the absence of any reference to religion in the UDHR, apart from the principle that it ought not to be used to discriminate against people and that the individual has the right to choose to be religious (or not). Human rights transcend particularities and are universal. By contrast, religious traditions often treat individuals differently (e.g. men and women; homosexuals and heterosexuals) and moreover can be manipulated to serve the interests of particular groups or individuals.

However, there are also very many examples of cases in which religious traditions are involved in different ways in support for human rights, for example,

Plate 5.1 *Buddhist monks taking part in anti-government protests in Myanmar, September 2007*

Source: racoles (Creative Commons). Online, available at:
http://commons.wikimedia.org/wiki/File:2007_Myanmar_ protests.jpg.

through involvement in human rights campaigns and advocacy (Plate 5.1). Religious traditions and organizations are often at the forefront of international human rights campaigns, with FBOs such as Christian Aid, World Vision or Islamic Relief expressing strong support for human rights as integral to their work as well as actively engaging in human rights campaigns within vulnerable locations such as Colombia, Sudan or the Democratic Republic of Congo. Many of these organizations also explicitly view development as a human right and consider that poverty exacerbates rights abuses, as well as itself being a symptom of rights abuses. As the head of policy and research at Islamic Relief stated during a debate about the relationship between poverty and human rights, which also involved the director of Amnesty International UK:

> Amnesty has been drawn into campaigning on poverty from a position of human rights, with Islamic Relief it has been the opposite. We traditionally work on poverty, and the links between poverty and human rights have driven us into the field of human rights.
>
> First, we came to realise that disasters are closely related to poverty. Disasters cause a great deal of suffering, but a much larger number of people suffer as

a result of poverty. When working in the field of poverty alleviation, we came to realise that our projects were good in and by themselves but that their impact would remain limited if a few root causes of poverty are not addressed. Human rights violations are among these root causes.[4]

While western human rights discourse, including that within mainstream development, has tended to ignore religion and culture, the examples in Box 5.2 suggest that the idea of the 'right to development' (RTD), for example, finds support within a range of different religious traditions (Plate 5.2).

Box 5.2

Religions and 'the right to development': some examples

Hinduism and the 'right to food'

Khare, in an article that discusses state-administered programmes to ensure 'food-for-all' in India, tells us that the notion of *sadharana dharma* (universal *dharma* or duty) was invoked by his informants as a basis for sharing food with the poor. Moreover, they argued that *sadharana dharma* should take precedence over the restrictive and exclusive rules of *jati dharma* (the duties followed by different castes) that might otherwise result in uneven food entitlements. While in practice people do not always uphold this aspect of *sadharana dharma*, he suggests that state food programmes are more likely to be successful if they are sensitive to the ways in which Hindu culture can both impede and support the goal of a hunger-free society. He draws our attention to the feeding of strangers during religious festivals 'irrespective of the condition and status of the recipient' (1998: 261). He argues that although 'religious charity alone could not wipe out hunger and poverty' (ibid.: 258), it is sociologically relevant to ask 'what keeps the traditional Indian food-gifting charity and philanthropic initiatives isolated from those that the state launches?' (ibid.: 258).

Pope Benedict XVI: *Caritas in Veritate* ('Charity in truth'), 2009

In his 2009 encyclical Pope Benedict XVI stated that:

11. . . .In the course of history, it was often maintained that the creation of institutions was sufficient to guarantee the fulfilment of humanity's right to development. Unfortunately, too much confidence was placed in those institutions, as if they were able to deliver the desired objective automatically. In reality, institutions by themselves are not enough, because integral human development is primarily a vocation, and therefore it involves a free assumption of responsibility in solidarity on the part of everyone. Moreover, such development requires a transcendent vision of the person, it needs God: without him, development is either denied, or entrusted exclusively to man [*sic*], who falls into the trap of thinking he can bring about his own salvation, and ends up promoting a dehumanized form of development. . . .

27. . . .The right to food, like the right to water, has an important place within the pursuit of other rights, beginning with the fundamental right to life. It is therefore necessary to cultivate a public conscience that considers *food and access to water as universal rights of all human beings, without distinction or discrimination.*[5]

Cairo Declaration on Human Rights in Islam, 5 August 1990

Article 17:

(a) Everyone shall have the right to live in a clean environment, away from vice and moral corruption, that would favour a healthy ethical development of his [*sic*] person and it is incumbent upon the State and society in general to afford that right.

(b) Everyone shall have the right to medical and social care, and to all public amenities provided by society and the State within the limits of their available resources.

(c) The States shall ensure the right of the individual to a decent living that may enable him to meet his requirements and those of his dependents, including food, clothing, housing, education, medical care and all other basic needs.[6]

Plate 5.2   *Volunteers at the Bangla Sahib Sikh Gurdwara, Delhi, prepare roti bread for the worshippers*

Source: Archer10 (Dennis Jarvis) (Creative Commons). Online, available at: http://en.wikipedia.org/wiki/File:Bangla_Sahib_Gurdwara_Delhi_-_Temple_and_Food.jpg.

The tradition of *langar*, or 'free kitchen', at Sikh gurdwaras, where food is served to all visitors for free, regardless of who they are, can be traced to the first Sikh guru, Guru Nanak, in the sixteenth century.

To redress the marginalization of religion from mainstream rights discourses, since the second half of the twentieth century numerous volumes have been published, outlining support for human rights principles across different religious traditions (e.g. Rouner 1994; Bloom *et al.* 1996; Ghanea 2009; Traer 1991), and international conventions have been held, bringing together different religious leaders to express the ways in which their traditions support human rights. The study by Traer, for instance, emphasizes not only the input from liberal Protestant groups into the formulation of the UDHR and the stress within recent Catholic theology upon the affirmation and defence of human rights, but also that other 'world religions' globally, can be interpreted to show strong support for human rights principles. He writes:

> Human rights are at the center of a global moral language that is being justified, elaborated, and advocated by members of different religious traditions and cultures. This is true not merely in the West but also in Africa and Asia. It is true not only in the First and Second Worlds, where liberal and socialist human rights theories have evolved, but in the Third World as well. Jews, Christians, Muslims, Hindus, Buddhists, and advocates of religious traditions indigenous to Africa and Asia fundamentally agree about human rights.
>
> (1991: 10)

This statement would seem to suggest that the answer to the question '*Are religious traditions compatible with human rights?*' is 'yes', and therefore to suggest that the marginalization of religions from debate and action around human rights in development is a product of secular liberal views about religion rather than the fact that they have nothing positive to contribute (Tomalin 2006a). However, in addressing the question '*Are religious traditions compatible with human rights*' a distinction needs to be made between what is rhetoric (e.g. the above quotation from Traer), what is based on analysis of religious teachings and what is based on actions (i.e. how people actually live out religious teachings in particular contexts). The very fact that religious traditions are not homogeneous and that they can be interpreted differently means that Traer's argument can only be partially true. To take it at face value would be to gloss over differences and tensions that need to be taken rather more seriously such that engagement between development and religions in pursuit of human rights can bring about positive social change.

For instance, the importance of dignity and human rights may well be at the forefront of Catholic ethics, therefore seeming to support Traer's argument, but it is also the case that the Catholic approach to abortion and contraception, which afford rights to the foetus, are at odds with views about women's reproductive rights as indicated in UN documents such as the Convention on the Elimination of all forms of Discrimination Against Women (CEDAW). Article 12 of CEDAW reads:

> States Parties shall take all appropriate measures to eliminate discrimination against women in the field of health care in order to ensure, on a basis of

equality of men and women, access to health care services, including those related to family planning.[7]

This has been widely interpreted to include abortion as family planning. On the basis of this example, the rhetorical statement that all religions support human rights needs to be treated with caution. Instead, we need to look at the way in which religious teachings are interpreted in different contexts, and the impact that these interpretations have upon people's actions. This typically varies from place to place.

The same is equally true of other traditions. If we consider Islamic statements on human rights, these are understood as being based on Shari'ah law. However, interpretations of Shari'ah extend from conservative versions, where the law is read to the letter, and men and women are afforded different rights, to more liberal versions where equality is emphasized. For instance, as Keddie notes, a modernist and reformist approach to the Qur'an (one of the sources of Shari'ah law) is that it can be reinterpreted in the light of modern concerns. She continues that:

> An allied argument is to stress the spirit of the Qur'an – to use the book title of the South Asian reformer Ameer Ali – and to say that the Qur'an is egalitarian (largely true) and favors human rights, and that these general principles were meant to be extended to women's rights. There is also extensive reinterpretation of particular verses and passages. The Qur'an in the same chapter says that men can marry up to four wives if they can treat the wives equally, and later that no matter how hard they try men will not be able to treat wives equally. Putting the two together, it is logically held by the modernists that the Qur'an was against polygamy, as the conditions it lays down as requirements for polygamy it then says are impossible to meet. More generally, various passages are interpreted to refer to male–female equality.
>
> (1994: 86)

This discussion suggests that religions intersect with human rights discourse in complex and often contradictory ways, and the rhetorical statement that religions support human rights may only be true in some contexts. At times, a religion may be supportive of the 'spirit' of human rights discourse but on some issues it may be at odds with it. While liberal versions of different religions would seem to overlap more closely with human rights discourse, conservative versions hold more tightly to literal interpretations of texts and traditions, often making it difficult to find common ground. These debates are further complicated by the view that human rights as enshrined in the UDHR reflect western individualistic values that are not universally recognized in some other religio-cultural value systems. The UDHR is sometimes considered to be a form of imperialism that seeks to impose and universalize western values.

The following sections unpack these tensions and suggest that two closely related issues need to be addressed. First, does the emphasis upon individual rights compromise religious understandings of the person and their duty to their religion as well as to other members of the community of believers or practitioners? Second, whereas mainstream human rights discourse views human rights as universal, religious versions of human rights are expressed in terms of particular teachings and will therefore vary from place to place. How can human rights debates, including those within development, accommodate the tension between universalism and cultural relativism?

Duties versus rights

From the point of view of members of religious traditions, there have been a number of criticisms of the Universal Declaration of Human Rights. As Little observes, 'heated debates have emerged over the nature, basis and scope of human rights' (1999: 153). With respect to the nature of human rights, as Traer writes, 'faith in human rights seems to undercut the sense of duty which is so much part of traditional culture in both its Eastern and Western forms' (1991: 5). Religious approaches to the rights debate typically stress the importance of duty and often maintain that an emphasis on individual rights compromises religious understandings about the nature of the person. Whereas the understanding of human rights as enshrined in the UDHR is that every human has equal rights because they are human and these cannot be taken away and should not be in any way compromised, religious traditions often take a more conditional view of human rights. For instance, according to many religious worldviews, the rights that a human being has are conditional to who they are (e.g. male or female) as well as how they act: in particular, humans have a duty to act in accordance with divinely sanctioned laws and expectations. This emphasis away from a focus on individual rights that pre-exist the individual can be contrasted with an emphasis in many religions that an individual is only afforded those rights if they undertake their duty in accordance with the teachings of their religion. Thus, at the risk of simplifying, an important difference between religious approaches to rights and secular liberal approaches is that for the former the interests of the faith as a whole and the community of believers come before the interests of the individual and their claim to rights. As Mayer writes, quoting an Iranian journalist commenting on Islam, '[T]he purpose of religion is to lead human beings to perfection; not everything can be permitted' (1994: 317).

Cultural relativism versus universalism

In addition to debates about the nature and basis of human rights, the universality of human rights has been repeatedly called into question. As Little argues, 'the underlying controversy is generally identified as a debate about cultural relativism' (1999: 153; see also Nussbaum 2000; Taylor 1999; Renteln 1990; Tilley 2000; Perry 1997). For example, critics argue that human rights discourse marginalizes people in the Global South because it fails to acknowledge the ways in which their cultural and religious traditions shape social ethics. In response to this critique there have been attempts to formulate human rights declarations that mirror different religious or cultural systems. Two examples will be discussed: the Bangkok Declaration on Human Rights (1993) and the Cairo Declaration on Human Rights in Islam (1990). Both declarations call the universality of human rights into question, since they offer *alternatives* to the UDHR that are shaped by different views about social ethics. The Bangkok Declaration emerges from the context of the so-called Asian values debate, whereas the Cairo Declaration takes as its starting point Islamic Shari'ah law. However, both declarations can also be viewed as attempts to resist western hegemony. From the point of view of advocates of the UDHR, this is a tricky position to be reconciled with, since attempts to renegotiate the full implications of the UDHR (i.e. the primacy of equal, inalienable and indivisible individual rights) are often perceived as paving the way for human rights abuses.

The Asian values debate

The Asian values debate highlights the tension between cultural relativism and universalism. Having its roots in the colonial period, it intensified in the 1990s through the support of particularly vocal advocates such as Mahathir Mohamad and Lee Kuan Yew, former prime ministers of Malaysia and Singapore (Hill 2000). It aimed to draw a distinction between western values and Asian values that could then be used to explain cultural and economic differences between different socio-political contexts (i.e. East and West). However, it also served as a means of distancing its proponents from western scrutiny where they felt that the West was dictating how they should behave through invoking human rights arguments, which they believed served western interests.

Peerenboom identifies two main dimensions to the debate. The first area of contention is related to human rights and the second to economic development (2003: 2–3). With respect to human rights, the main issue relates to the tension between universalism and relativism, specifically the extent to which Confucianism, Buddhism, Islam and Hinduism are compatible with human rights and democracy. It is argued that western values, while portrayed as universal, are actually a cultural product and that, because of cultural differences, 'values' in Asia are different to those in the West. Within this discourse, western values are associated with individualism and materialism, whereas Asian values are depicted as prioritizing respect for community and authority (Kausikan 1995–1996, 1997; Chan 1995). Therefore, Asian values are considered to provide support for duties rather than rights, and to lend themselves to the emergence not of democracy but, instead, of single-party authoritarian politics. From the point of view of economic development, the Asian values debate has a special significance, since recourse to Asian values was also used to account for the success of the so-called tiger economies of East and South-East Asia, at least until the financial crisis of the late 1990s. According to Peerenboom, the debate questioned 'whether authoritarian or democratic regimes are better able to achieve sustained economic growth and whether Asian versions of capitalism are superior to the varieties of capitalism found in Western liberal democracies' (2003: 3). As Sen, a critic of the Asian values argument, also points out, it 'argues for authoritarian governance in the interest of economic development' (1997). Underpinning the Asian values argument is the view that economic, social and cultural rights, particularly the 'right to development', are to be promoted above civil and political rights.

However, critics of this position, of which there have been many, are concerned that the 'insistence on duties may disguise unwillingness to acknowledge the well-founded claims of citizens to respect for their internationally recognized human rights' (Marks 2001: 11). For instance, as Marks writes, Kim Dae Jung, a former human rights activist, and president of South Korea between 1998 and 2003,

> rejected the 'Asian values' argument, which he found 'not only insupportable but self-serving.' Kim considered that '[t]he biggest obstacle [to establishing democracy and strengthening human rights in Asia] is not its cultural heritage but the resistance of authoritarian rulers and their apologists.'
>
> (2001: 11; Jung 1994)

Moreover, whether there is indeed a distinct set of Asian values that are held across Asia has been questioned. Support for the debate is found mainly in East and South-East Asia, with the concept largely incorporating Confucian values but also extended to include Asian values more broadly, such as those influenced by Hinduism, Buddhism and Islam.

A legacy of this debate is the Bangkok Declaration on Human Rights. A few months before the World Conference on Human Rights in Vienna in 1993, a regional meeting for Asia (although not all participants were from Asian nations) was held in Bangkok. Participants argued that the UDHR had been established without the participation of 'Asian' states and that it needed to be updated and revised to reflect their input. (Waltz 2001, 2002, however, challenges the legitimacy of this view.) As Lindsey writes:

> The Bangkok Declaration was the high tide of the 'Asian values' argument, at least at the formal policy level. Whatever the niceties of its phrasing, the instrument was widely understood as based on the argument that Asians shared distinct values that were incompatible with values shared by Westerners and that therefore the West should not rely on its construction of human rights to intervene in affairs of Asian states.
>
> (2004: 286)

Article 8 states that:

> [w]hile human rights are universal in nature, they must be considered in the context of a dynamic and evolving process of international norm-setting, bearing in mind the significance of national and regional particularities and various historical, cultural and religious backgrounds.[8]

Thus, while the universality of human rights are affirmed, these words were widely interpreted as simultaneously undermining the universality of human rights by suggesting that they were culturally relative. In particular, it suggested the view that the UDHR was based upon western understandings of individual rights that do not necessarily reflect traditional 'Asian values' (Cumaraswamy 1997; Jacobsen and Bruun 2000; Peerenboom 2000, 2003; Bauer and Bell 1999). Moreover, while the Bangkok Declaration seemed to take the Asian values debate in a new direction in affirming the indivisibility of economic, social, cultural, political and civil (ESCPC) rights (article 10), overall it appears to prioritize economic rights, asserting the importance of the right to development (articles 17 and 18).

Whether the somewhat contradictory nature of the document is a product of its actual content or the way that it has been read by those critical of its underlying motives is debatable. Not surprisingly, it has been criticized for only 'paying lip-service to widely accepted principles of human rights' (Lee 1998: 1; Tang 1995) and resulting in a situation where 'a government might claim that its traditions, customs or conditions – for example, religious beliefs, traditional notions of the role of women or even the threat of subversion – permitted it to restrict or abuse the rights of its citizens' (Lee 1998: 1). Such warnings may not be unfounded, but the suggestion that the Asian values debate is at least partially understood as an attempt to assert power in an essentially hegemonic discourse might indicate the futility of attempting to trump 'Asian values' with 'human rights'. It was never formally adopted by the United Nations, but is a reflection of a debate that has as much to do with the attempt to resist western hegemony as it does with cultural relativism. For instance, the document is quite justified in challenging the western champions of human rights for their perceived 'double standards': for not consistently pursuing human rights causes when to do so compromises their interests; for not being honest about human rights problems in their own countries; and for denying responsibility 'for a colonial legacy of rights abuses in the Asian countries over which they now sat in judgment' (Peerenboom 2000: 296). However, the moral force of this critique is somewhat lessened since its originators are themselves critiqued for promoting Asian values as a mask to hide human rights abuses.

It may be difficult to completely resolve this argument, since the idea of distinct differences between Asian and European values is strongly held by many, and moreover it is a political or ideological view that is 'often associated with the need to resist Western hegemony' (Sen 1997). While the Asian values debate often tends to overstate the case, ending up essentializing both Asian and western cultures, opponents of the debate have also been guilty of the same. As is outlined in Box 5.3, a good example of the pervasiveness of this tendency to polarize cultural difference between the East and the West is Samuel P. Huntington's 'clash of civilizations' thesis (1993), which has received criticism from commentators such as the postcolonial theorist Edward Said for essentializing and homogenizing non-western cultures.

Box 5.3

The 'clash of civilizations' thesis

According to Huntington, the current phase of world politics is dominated by clashes over culture, and he argues that western concepts 'differ fundamentally from those prevalent in other civilizations' (1993: 40). Among western concepts he lists 'ideas of individualism, liberalism, constitutionalism, human rights, equality, liberty, the rule of law, democracy, free markets, the separation of church and state' (ibid.: 40) and states that they 'often have little resonance in Islamic, Confucian, Japanese, Hindu, Buddhist or Orthodox cultures' (ibid.: 40). In particular, he posits an Islamic civilization that is opposed to the West and argues that the promotion of universal values is a waste of time, being likely to provoke backlash.

Edward Said, the postcolonial critic, was one among many vocal critics of Huntington's thesis. In an article entitled 'Clash of Ignorance' (2001), he wrote:

> [T]he personification of enormous entities called 'the West' and 'Islam' is recklessly affirmed, as if hugely complicated matters like identity and culture existed in a cartoonlike world where Popeye and Bluto bash each other mercilessly, with one always more virtuous pugilist getting the upper hand over his adversary.
>
> (Said 2001)

And elsewhere:

> My concern . . . is that the mere use of the label 'Islam,' either to explain or indiscriminately condemn 'Islam,' actually ends up becoming a form of attack . . .'Islam' defines a relatively small proportion of what actually takes place in the Islamic world, which numbers a billion people, and includes dozens of countries, societies, traditions, languages, and, of course, an infinite number of different experiences. It is simply false to try to trace all this back to something called 'Islam,' no matter how vociferously polemical Orientalists . . . insisted that Islam regulates Islamic societies from top to bottom, that *dar al-Islam* is a single, coherent entity, that church and state are really one in Islam, and so forth.
>
> (Said 1997: xv–xvi)

Peerenboom (2003) suggests, however, that the overly aggressive and polemical nature of the first phase of the Asian values debate had receded by the time of the financial crisis in 1997. Since then, the debate 'has been much less politicized and the arguments more sophisticated and balanced' (ibid.: 52). There has been an 'attempt to move beyond debates about universalism versus relativism' (ibid.: 56) through emphasizing similarities or overlaps between eastern and western cultures, where human rights and their underlying values such as tolerance and freedom are seen to have foundations across different religious and cultural traditions, rather than emerging as a product of modern western secular liberalism. As Sen (1997) argues:

> Our ideas of political and personal rights have taken their particular form relatively recently, and it is hard to see them as 'traditional' commitments of

Western cultures. There are important antecedents of those commitments, but those antecedents can be found plentifully in Asian cultures as well as Western cultures.

One way of approaching this has been the idea of 'overlapping consensus', based upon the moral philosophy of the North American philosopher John Rawls (Rawls 1996; Taylor 1999; Renteln 1990; Tomalin 2006a). This points to a view of 'universalism' that differs in important ways from that underpinning the UDHR. We will return to this later in the chapter.

The Cairo Declaration on Human Rights in Islam

Before looking at attempts to move beyond the universalism versus relativism debate and the implications for development, we will examine another alternative to the UDHR, this time emerging from within Islam. The Cairo Declaration on Human Rights in Islam was adopted by fifty member states of the Organization of the Islamic Conference (OIC) in August 1990[9] and states that 'the Islamic Shari'ah (Islamic law) is the only source of reference for the explanation or clarification of any of the articles of this Declaration' (Mayer 1994: 329).

Ann Elizabeth Mayer, a North American law professor who has written widely on Islam and human rights, has discussed the background to the Cairo Declaration in a detailed and lengthy article (1994). She notes that after the Second World War, Muslim governments began to define their position vis-à-vis international human rights and at that time seemed to accept the international consensus. Saudi Arabia was the only Muslim country in the United Nations that failed to vote for the UDHR on the basis that it did not fit with Islamic teachings. By the 1970s, however, the view that human rights norms were incompatible with Islam was heard more frequently, and after the 1979 revolution in Iran, Iran became a strong supporter of this view. In 1983 Iran's UN ambassador, Sai'id Raja'i Khorasani, depicted the UDHR as based on a 'secular understanding of the Judeo-Christian tradition, which could not be implemented by Muslims and did not accord with the system of values recognized by the Islamic Republic of Iran' (ibid.: 315–316). He suggested that his country would willingly violate it, as this amounted to a choice between 'violating the divine law of the country and violating secular conventions' (ibid.: 316). Mayer writes that this 'trend to plead Islamic particularism culminated in 1990, when Muslim countries joined in putting forward the Cairo Declaration of Human Rights in Islam' (ibid.: 322). As with the Asian values debate, the main conflict was over political and civil rights, and particular departures from the UDHR arose with respect to religious freedom and equality, particularly of women and non-Muslims.

A meeting to prepare the OIC document was held in Tehran in December 1989. Mayer argues that the document is a hybrid, drawing on elements of international law, Islamic law and the preferences of its originators, and that it sought to restrict political and civil rights in the name of Shari'ah but 'without any attempt at defining what limits the shari'a would entail' (1994: 329). This lack of clarification exists because there is 'no settled jurisprudence on the question of how reference to overriding Islamic criteria should affect modern rights norms' (ibid.: 329). Mayer carefully explores ten areas relevant to human rights considerations and demonstrates that their provision in the Cairo Declaration falls well below that of the UDHR. The areas she explores are equality for women, freedom of religion, freedom of the press, freedom of assembly and association, democratic freedoms, criminal justice, states of emergency, the prohibition of genocide, the right to security and privacy, and the prohibition of colonialism and slavery. Writing three years after the declaration was formulated, she notes that in actual fact it had had little impact on how Islamic governments dealt with rights and that the rights enshrined in state constitutions, often also based on Islam, were typically better than those expressed in the Cairo Declaration (ibid.: 350). Moreover, she argues that it is important to draw a distinction between the 'official positions' on Islamic human rights and those held by NGOs, many of which do not see a contradiction between human rights and Islam, suggesting that the conflict is not about Islam per se but rather about politics. Box 5.4 provides some background to the work of one of the best-known examples of human rights advocacy within Islam, the Malaysian organization Sisters in Islam, which describes itself as

> a group of Muslim professional women committed to promoting the rights of women within the framework of Islam . . . based on the principles of equality, justice and freedom enjoined by the Qur'an as made evident during our study of the holy text.[10]

Box 5.4

Sisters in Islam

'*If God is just as Islam is just why do laws and policies made in the name of Islam create injustice?* This was the burning question faced by the founding members of Sisters in Islam (SIS) when they began their search for solutions to the problem of discrimination against Muslim women in the name of Islam' (emphasis added).

Sisters in Islam was founded in Malaysia in 1987, by women lawyers and their friends, 'to study problems associated with the implementation of new Islamic Family Laws that had been legislated in 1984, and enforced in 1987'. In addition to law reform, however, members of the group increasingly felt that 'dealing with the law alone was insufficient, especially in view of the fact that Islam was being increasingly referred to as a source of injustice and oppression'. Thus, 'some within the group felt the urgent need to re-read the Qur'an to discover if the Text truly supported the oppression and ill-treatment of women'. With respect to passages that appeared to legitimize the oppression of women, the organization was able to provide interpretations and explanations, suggesting that practices such as domestic violence or polygamy (unless all wives could be treated equally) were not consistent with Islam.

Source: www.sistersinislam.org.my/page.php?35 (last accessed 10 September 2012).

Therefore, while the Cairo Declaration professes to present an Islamic approach to human rights, it does not achieve its aim, since there is no such thing as a monolithic Islam recognized across Muslim countries that shapes in a uniform fashion domestic approaches to rights as well as responses to international norms. Mayer concludes that

> [t]he constructs of Islamic rights in the civil and political sphere, that one finds in schemes like the Cairo Declaration and the Saudi Basic Law, are designed to shore up the political interests of those promoting them and have only a tenuous connection to Islamic culture. They borrow extensively from Western rights models and mine the Islamic heritage only very selectively – shutting out the enlightened, modern perspectives of Muslims who are supportive of human rights.
>
> (1994: 402–403)

* * *

The existence of these alternatives to the UDHR from different religio-cultural traditions might seem to support Huntington's thesis that there is a 'clash of civilizations', and that attempts to further assert the universality of western human rights will only serve to create hostile backlash. However, as Mayer and others have demonstrated, many people in the Global South do support human rights ideals, although they resent the way in which the West demonstrates double standards in endorsing universality while failing to apply rights principles in practice. At the grassroots level of human rights activism there seems to be rather more of an 'overlapping consensus' with the principles of universal human rights than these two alternatives to the UDHR suggest.

Moving beyond universalism versus relativism? 'Overlapping consensus' and human rights

Academic reflection upon both the Asian values debate and the Cairo Declaration expresses suspicion about the motives of those arguing that western human rights are at odds with those found in other cultures (Kausikan 1995–1996, 1997; Chan 1995; Mayer 1994). However, it is also the case that the way in which western human rights discourse has tended to portray other value systems as homogeneous *and* oppositional to rights, therefore firming up the claim that it is necessary to have a universally recognized system of rights, also fails to capture the way in which rights are enacted in practice. Both views (universalist and relativist) arguably belie political interests and are essentialist positions. Another approach that has become popular in rights debates is the attempt to reach an 'overlapping consensus' 'among divergent, and irreducibly particularistic, cultural and religious beliefs' (Little 1999: 159). While some interpretations of religious traditions, particularly those that are 'conservative' or 'fundamentalist', are at odds with many human rights principles, more liberal versions of traditions typically argue that there is much overlap between their approach to human rights and that of mainstream rights discourse, and that their marginalization from the debate reflects secular liberal attitudes towards religion as outdated and backward-looking rather than the fact that they have nothing positive to offer.

One important contribution that many consider religious traditions have made to rights discourses is to rebalance an emphasis upon individualistic rights with recognition of corresponding duties. This suggestion that modern human rights discourse has sidelined duty and responsibility is not confined to religious versions of rights (Laws 2003), but faith traditions typically argue that they have a unique contribution to make towards human rights thinking in this respect (Kung and Kuschel 1993; Sharma 1999: 539; Runzo *et al.* 2003). The philosopher Onora O'Neill, for instance, introduced this discussion into her 2002 Reith Lectures, arguing that 'Declarations of Rights ostensibly offer something to everybody, but they do it without coming clean about the costs and demands of respecting the rights they proclaim' (2002: 3). A document entitled 'A Universal Declaration on Human Rights by the World's Religions' (Sharma 1999: 539; Runzo *et al.* 2003) reflects this approach. The document was drafted by academics with the aim of inviting religious communities to respond. It offers a revised version of the UDHR, with a significant change being the incorporation of the idea of 'duty' into each article of the original declaration. Thus, article 18 is embellished with the sentence: 'Everyone has the duty to promote peace and tolerance among different religions and ideologies' (Sharma 1999: 542).

Supporters of the 'overlapping consensus' approach stress the importance of expressing and explaining the content of human rights with respect to different traditions, rather than purely within the framework of secular liberalism (Bauer and Bell 1999; An-Na'im 1992; Bloom *et al.* 1996; Sharma 2004; Little 1999). Sen argues against the idea that universal human rights only have western roots. Instead, in a critique of the Asian values debate, he points out that support for freedom, tolerance and equality can be found within a broad spectrum of Asian traditions where 'the roots of modern democratic and liberal ideas can be sought in terms of constitutive elements, rather than as a whole' (1997: 16). While he concedes that these values often existed side by side with their opposites, and were in general not applied equally to women or slaves, the point is that freedom, tolerance and equality are not a western invention. Both Sen and Nussbaum, for example, point to the example of the Indian king Ashoka (third century BCE), a convert to Buddhism who invoked Buddhist teachings in support of religious tolerance (Sen 1997; Nussbaum 2000). As Sen (1996) writes:

> If the grabbing of 'Asian values' by the champions of authoritarianism has to be effectively and fairly questioned, what is needed is not the claim – often implicit– of the superiority of what are taken as Western values, but a broader historical study of Sanskrit, Pali, Chinese, Arabic, and other Asian literatures (in relation to corresponding writings in the Western classics). And nearer our times, acknowledgment would have to be made to the contributions of national leaders such as Mahatma Gandhi or Dr. Sun Yat-sen, who were, already a

hundred years ago, cogently vocal in defense of the widest forms of democracy and political and civil rights.

The problem highlighted by the above critiques (both those which reject western human rights, such as the typical Asian values debate, and the more moderate perspective articulated by Sen) suggests the perception that human rights discourse is a 'comprehensive doctrine' (Rawls 1996). However, drawing on the thought of the philosopher John Rawls, public dialogue in a pluralistic global society is arguably more effective when it engages with 'overlapping consensus' rather than 'comprehensive doctrines' (ibid.). With respect to human rights, this does not mean abandoning the search for universals, only that they should be convergent universal norms that can be justified from different religious and philosophical perspectives. As Taylor writes, many critics of human rights discourse are in fact 'eager to espouse some universal norms, but they are made uneasy by the underlying philosophy of the human person in society . . . [which] seems to give pride of place to autonomous individuals, determined to demand their rights' (1999: 128). He questions, for example, the extent to which this focus on the individual fits with the 'Confucian emphasis upon close personal relationships, not only as highly valued in themselves, but as a model for society . . . [or the] Theravada Buddhist search for selflessness, for self-giving and *dana* (generosity)' (ibid.: 128–129).

Thus, underpinning the overlapping consensus approach is the idea of anchoring human rights within a set of cross-cultural standards, thereby rendering them 'more likely to be accepted and taken seriously' (Renteln 1990: 138). Questions about who would decide upon such cross-cultural universals, and whether a meaningful consensus could ever be reached, do continue to occupy those engaged in research in this area. However, as a broad principle to guide rights-based work in development there is much to recommend the overlapping consensus approach. It recognizes not only that human rights principles can be found across different religio-cultural traditions but also that this approach is likely to be more strategically effective in enabling the realization of rights in different contexts and debunking the idea that rights and religion are intrinsically at odds.

Summing up

The aim of this chapter was to address the following questions: *Are religious teachings compatible with human rights? What are the implications for 'development'?* The above discussion has demonstrated that there are a range of relationships between religions and human rights. Religious traditions can act against human rights, generate alternative versions of human rights and also directly support modern human rights frameworks, as illustrated by the overlapping consensus approach to rights. The diverse manifestations of religious traditions as they are lived out in different contexts mean that the tendency within mainstream development policy, in particular, to view human rights and religion as invariably at odds needs to be challenged. While, as Balchin (2011, 2012) suggests, the *actual practice* of development on the ground has tended in many cases to involve greater awareness of the religious implications for human rights as well as negotiation with religions in rights-based work, this has rarely filtered back to formal policy formulation and the literature produced by development organizations and agencies. She suggests instead that 'rights and religion (or rights and culture) need not be seen as opposites that need balancing which implies that one or the other has [been] compromised in the process of development' (2011: 25).

Consideration of religion in terms of how it can hinder as well as support rights is relevant to the pursuit of human rights in development and can play a role in bridging the chasm between the theory and realization of rights. However, a number of steps need to be taken in order for this to become a reality. There is a need for a much greater emphasis within development studies upon research to find places where human rights might intersect with other moral discourses and how such an emphasis can enhance the pursuit of progressive development goals in particular locations. However, while this type of exercise should draw upon insights from different academic disciplines, in order for it to be of relevance to the development process it would need to engage with the reality of the lives of those who are living in poverty or who are marginalized and oppressed in specific locations; it cannot be a theoretical and purely theological exercise. In addition, development actors need to acquire better understandings of the ways in which religious traditions might block human rights in different contexts and how alliances might be sought with more liberal religious actors to counter conservative or fundamentalist views. In contexts where people's rights are impeded by religion and/or where religions are socially powerful, to engage with religious leaders and to refer to textual traditions, perhaps via 'faith-based organizations', is one way of educating people about the rights that they have (according to their religion, but also as set out in the UDHR), as well as establishing a powerful and persuasive lobby against interpretations of religious traditions that are oppressive and marginalizing. One organization that has a strong track record in this approach is the Asia Foundation, particularly in engaging with religious leaders within women's empowerment projects. For instance, its project Advancing Human Rights and Women's Rights within an Islamic Framework across South Asia is described as

> [p]romoting gender equity and women's rights within an Islamic framework in South Asia, as part of a regional program across Afghanistan, Bangladesh, and Pakistan through: developing a training curriculum for Muslim leaders on women's human rights; fostering an exchange of ideas and discourse in South Asia on the relationship of Islam, human rights, and women's rights; creating a locally appropriate training curriculum for religious leaders in South Asia to advance social justice for women; and expanding and strengthening regional networks of Muslim scholars and leaders working to advance women's rights in the context of Islam.[11]

It is notable that many of the examples of development projects that engage with religion around securing rights are concerned with women's rights and gender equality. In the case of women's rights, which are addressed more specifically in the following chapter, approaching rights through a religious framework can also assist in preventing dangerous backlash from fundamentalist or conservative groups. The promotion of rights in many countries runs the risk of backlash, particularly where women's rights are concerned, with rising forms of fundamentalism and extremism seeking to control and limit the gains that have been made at both local and international levels.

Questions for discussion

1 Discuss the extent to which religious traditions can be considered to be compatible with human rights.

2 Why have religious traditions tended to be marginalized from mainstream human rights debates, including those within donor-driven development?

3 What is the significance of the 'universalism' versus 'cultural relativism' debate for a consideration of religions and human rights?

4 In what ways is the 'overlapping consensus' approach to human rights relevant for 'human rights-based approaches' to development?

Recommended reading

Banchoff, Thomas and Robert Wuthnow (eds) (2011) *Religion and the Global Politics of Human Rights*, New York: Oxford University Press. This collection of essays covers many of the themes raised in this chapter, including the tensions between universal and relative understandings of human rights and the extent to which modern western rights regimes are imperialist.

Ghanea, Nazila (ed.) (2010) *Religion and Human Rights* (4 vols), London: Routledge. These volumes contain essays examining the background, history and nature of human rights. They explore rights issues within Islam, Christianity, Judaism and African religions.

The following texts are examples of the many books that explore the approaches to human rights within the 'world religions':

Keown, Damien, Charles S. Prebish and Wayne Rollen Husted (eds) (1998) *Buddhism and Human Rights*, Richmond, Surrey, UK: Curzon Press.

Sharma, Arvind (2010) *Hindu Narratives on Human Rights*, Santa Barbara, CA: Greenwood.

Witte, John Jr and Frank S. Alexander (eds) (2011) *Christianity and Human Rights: An Introduction*, Cambridge: Cambridge University Press.

Websites

http://religionhumanrights.com This website, developed by the US academic Robert Traer, contains versions of material previously published by Traer on a wide range of religion and rights topics.

www1.umn.edu/humanrts/edumat/studyguides/religion.html This website provides a study guide, 'Freedom of Religion or Belief', from the University of Minnesota Human Rights Library.

www.christianaid.org.uk/whatwedo/in-focus/human-rights-christian-aid/index.aspx This section of the Christian Aid website has links to information about its current human rights work.

www.ihrc.org This is the website of the Islamic Human Rights Commission. It is an independent, not-for-profit, campaign, research and advocacy organization based in London, and has consultative status with the United Nations Economic and Social Council.

www.law.emory.edu/ihr/acessay.html This site has links to articles on Islam, human rights and related topics. It is hosted by the Law Program at Emory University.

www.religlaw.org This is the website of the Religion and Law Consortium, 'a research forum for legal developments on international law and religion or belief topics'.

6 Gender, religions and development

This chapter will:

- address the following question: *Is religion bad for women's development and the pursuit of gender equality?*;
- examine how 'gender and development' (GAD) has approached religions;
- discuss the ways in which attitudes towards women in religious traditions present challenges for mainstream development policy and practice;
- look at the engagement with 'religious feminisms' as a culturally located strategy for encouraging and realizing female empowerment;
- evaluate concerns that the 'turn to religion' by donor-driven development policy and practice has gender costs that have not been adequately considered.

Introduction

In this chapter the following question is addressed: *Is religion bad for women's development and the pursuit of gender equality?* Until recently, 'gender and development' (GAD) programmes within mainstream development agencies and organizations have tended to more or less ignore or reject religion as either a relevant or a potentially positive influence in women's lives and instead have drawn upon secular styles of feminism that typically consider religion to be backward, primitive and ultimately against women's best interests. A recent study by Seguino, which relates cross-country data on gender attitudes from the World Values Survey to levels of religiosity, strongly supports the view not only that religiosity 'is positively associated with gender inequitable attitudes' (2011: 1310) but also 'that the effect of religiosity extends beyond attitudes to negatively impact several measures of gendered well-being outcomes' (ibid.: 1317). While it is impossible for a large-scale survey to adequately capture the diversity within religions as well as between religions, this study does demonstrate the same patterns even once GDP and level of development were controlled for (ibid.: 1317), and also that 'no major religious denomination stands out as being significantly more strongly associated with gender inequitable attitudes than the others' (ibid.: 1313).

Nonetheless, while religion certainly does legitimize values and rules that disempower women, the importance of religion in the lives of millions of poor women across the globe means that secular feminism is often perceived by them not only as western but also as lacking cultural relevance. Rather than rejecting religion for its inherent patriarchy, styles of 'religious feminism' have emerged across the globe. These argue for reinterpretations of religious systems that are consistent with the 'core' values of the particular religious tradition as well as various types of feminist thinking. Such a strategy is attractive to women who wish to employ a religious narrative to guide their politics of empowerment rather than relying on the secular rhetoric of mainstream (western) feminist discourses; but it can also be expedient to employ a religio-cultural narrative in contexts where religion continues to influence women's life chances.

Over the past decade the 'negative' and neglectful attitude towards religion in donor-driven development processes has begun to recede, and religious issues are receiving more consideration, including the direct engagement of donors with 'faith-based organizations' (see Chapter 8) and religious leaders. An understanding of the influence that religious traditions have on women's social status or economic opportunity is increasingly being recognized as an important factor in the pursuit of female empowerment, both positively and negatively. Moreover, there are now a wider range of GAD-related programmes that themselves aim to directly engage with religion through a consideration of the role that 'religious feminisms' can play in women's empowerment, for instance in forming partnerships with organizations that pursue women's rights within a religious framework (Bradley 2006, 2010; Tadros 2010, 2011).

Nonetheless, despite the 'turn to religion' by donor-driven development (including within 'gender and development'), which many consider to be positive since it recognizes that religion is often important in women's lives, others are much more cautious (Tadros 2010; Balchin 2011). They argue that little attention has been paid to the gender consequences of engaging with religion, despite the fact that religions often support inequality between men and women, and are typically led by men, and a 'faith-based development' model may have gender consequences that have not been adequately considered. Seguino, for instance, strongly emphasizes the need to scrutinize the 'impact on gender equality of aid funneled through religious organizations' since 'donors may find that religious non-governmental organizations have a weaker record in improving women's relative well-being than non-religious organizations' (2011: 1317). There is some concern that many development actors now seem to be in rather a hurry to engage with religion (after decades of consciously avoiding it) and at times prioritize religious organizations in certain funding streams, or actively seek to support their work above that of secular actors. As Pearson and Tomalin have noted, 'there is an anxiety that in the rush to engage with a hitherto neglected group of stakeholders, the painful journey over the last 35 years to mainstream development gender equity objectives into overall development strategy is being sidelined' (2008: 47). Box 6.1 outlines what for many has been one of the most worrying threats for gender equality of the 'rush to find the religious' within donor-led development, namely the preference shown by the government of George W. Bush in the United States (between 2001 and 2009) for selecting FBOs as priority partners, particularly with respect to AIDS relief work.

Box 6.1

PEPFAR, FBOs and the impact on women

Former US president George W. Bush, in his January 2003 State of the Union Address, promised US$15 billion for the 'President's Emergency Plan for AIDS Relief' (PEPFAR). By 2005 a quarter of all PEPFAR partners were FBOs. This is a trend that reflects PEPFAR's explicit position that '[f]aith-based groups are priority local partners. . . . In certain nations, upwards of 50 percent of health services are provided through faith-based institutions, making them crucial delivery points for HIV/AIDS information and services' (PEPFAR 2005: 2).

The fact that 20 per cent of these funds were for HIV prevention, of which at least a third was to be spent on 'abstinence until marriage' programmes (which are more likely to be delivered by FBOs), has gender consequences. Women in developing countries have a higher incidence of HIV/AIDS than men and are likely to benefit from the promotion of consistent condom use (UNAIDS 2010).

In order to examine the extent to which religion is bad for women's development and the pursuit of gender equality, the chapter will begin with a discussion of the emergence of a 'gender and development' (GAD) approach within donor-driven development agencies and organizations. This will include an examination of the failure to seriously engage with the impact of religious and cultural diversity on women's lives, including engaging with various styles of 'religious feminism'. I will argue that approaches to women's empowerment that aim to address the structural reasons for women's disadvantage can benefit from considering religion as a relevant factor. However, what does it mean for GAD to 'engage' with religion and what are the particular challenges and opportunities likely to be encountered? These are the questions I will address in this chapter. First, we will explore the ways in which attitudes towards women in religious traditions present challenges for mainstream development policy and practice; second, we will look at the engagement with 'religious feminisms' as a culturally located strategy for encouraging and realizing female empowerment; and finally, we will return to the concerns outlined above: that the 'turn to religion' by donor-driven development policy and practice has gender costs that have not been adequately considered.

Gender studies and women's development

Work in the area of 'gender and development' is primarily concerned with issues such as the social differences between men and women, the economic marginalization of women, or violence against women as a product of gender bias. However, studies in this field have, on the whole, not taken the impact of religious values, beliefs and organizations upon gender relations that seriously.

Reflecting the dominant development paradigm, GAD approaches are typically both materialist and secularist. Religion is often cited as an impediment to women's access to employment, health care or education, but the dynamics of this interaction less often become the research focus. In the mainstream literature there is a failure to engage with the ways in which some feminists across the globe seek to transform gender relations through engaging with feminist interpretations of their traditions, as well as a lack of interest in the ways in which religion may inform alternative, sometimes more culturally appropriate, understandings of development.

In order to understand the neglect of religion within 'gender and development', we need to go back to the early 1970s. Following the publication of Ester Boserup's book *Women's Role in Economic Development* (1970), we find greater attention to the recognition that women are central to the development process and that without considering the subordinate position of women, poor countries were likely to remain poor (Momsen 2010: 11ff.). Considerations of religion and culture were absent in this analysis, reflecting the secularist under-pinnings of the women's movement, as well as a lack of emphasis (certainly at this early stage of feminist activism) of the fact that the challenges that 'First World' and 'Third World' women faced were likely to differ in important ways (Rathgeber 1990). Moreover, early 'women in development' (WID) approaches, which aimed to bring women into the development process and to ensure that they benefited from it, were criticized for failing to tackle the underlying structural reasons for gender oppression. Critics argued that WID relied upon the assumption that increasing women's opportunities (e.g. educational or employment) would be sufficient to end gender discrimination (for a discussion of the limitations of WID, see Pearson 2000; Jackson and Pearson 1998; Moser 1993; Rathgeber 1990). Neither, they argued, did it challenge the western and gender bias inherent within the development project itself.

By the 1980s we find a shift in focus to thinking about 'gender and develop-ment' (GAD). In contrast to earlier approaches, which just focused upon women, GAD also brought analysis of men's situations and roles into the picture. While there continued to be an emphasis upon understanding women's gendered disadvantage (a position also adopted in this chapter), there was an increasing realization that this could not be achieved without investigating various aspects of the construction of masculinities as crucial to understanding gender relations more broadly within developing contexts and also directly addressing men's attitudes and values in attempting to improve women's lives (e.g. Cornwall and White 2000; Cornwall *et al.* 2011). Moreover, as Moser writes:

> The WID approach . . . is based upon the underlying rationale that development processes would proceed much better if women were fully incorporated into them (instead of being left to use their time 'unproductively'). It focuses mainly on women in isolation, promoting measures such as access to credit and employment as the means by which women can be better integrated into the

development process. In contrast, the GAD approach maintains that to focus on women in isolation is to ignore the real problem, which remains their subordinate status to men. In insisting that women cannot be viewed in isolation, it emphasizes a focus on gender relations, when designing measures to 'help' women in the development process.

(1993: 3)

Moser (1993), drawing upon the earlier work of Molyneux (1985), makes a distinction between women's 'practical' and 'strategic' gender needs to argue that while development ought to cater for practical needs, such as access to employment, education or health care, this alone will not necessarily enable women to enhance their strategic position within the gender hierarchy. Thus,

> [s]trategic gender needs are the needs women identify because of their subordinate position to men in their society. . . .They relate to gender divisions of labour, power and control and may include such issues as legal rights, domestic violence, equal wages and women's control over their bodies.
>
> (Moser 1993: 39)

By contrast,

> [p]ractical gender needs do not challenge the gender divisions of labour or women's subordinate position in society. . . . Practical gender needs are a response to immediate perceived necessity . . . and often are concerned with inadequacies in living conditions such as water provision, health care and employment.
>
> (ibid.: 40)

By the 1990s it had become common to talk about 'gender mainstreaming' (Baden and Goetz 1998; Jahan 1995). As Baden and Goetz write, 'mainstreaming signifies a push towards systematic procedures and mechanisms within organisations – particularly government and public institutions – for explicitly taking account of gender issues at all stages of policy-making and programme design and implementation' (1998: 20). Women-only projects are not necessarily to be abandoned, but gender analysis should become integral to all stages of development processes, whether they are focused on women or not. Moreover, thinking about *gender* rather than *women* also implies the need to look at men's needs and roles, and how these are shaped in conjunction with those of women. An important aspect of gender mainstreaming has been the emergence of gender planning and training methods that aim to ensure gender sensitivity within policy and projects (Moser 1993; Macdonald 1994; Seed 1999; Kabeer and Subrahmanian 2000). While we find that all the major development agencies in the Global North have taken on board the importance of gender analysis (e.g. DFID 2008; World Bank 2011; UNDP 2011), others have argued that this widespread adoption of GAD has resulted in a watering down of its transformative potential: it becomes another box to tick rather than being a position that is deeply held and understood by individuals assigned to GAD work. As Pearson writes:

> It is widely accepted in these times that development must be informed by gender analysis and that particular attention must be paid to the needs of poor women – so much so that such positions have become commonplace rather than radical; indeed, many would argue that the ways in which gender matters have been integrated into development thinking and practice indicate a high degree of co-option of politicized feminist objectives rather than their success in transforming the development agenda.
>
> (Pearson 2000: 383; see also Smyth 1998)

Thus, the extent to which much GAD work has the potential to empower women might be questioned. One area that has received attention concerns the extent to which GAD is informed by western secular feminist models that fail to account for the different ways in which culture and religion shape women's oppression as well as the strategies they employ to deal with it. The context outlined above has not tended to foster a research agenda that is concerned with the impact of religion and culture upon women's lives in poor countries. WID approaches were more concerned with increasing women's access to resources, as a remedy to women's subordination, and GAD tends to focus upon the changes needed to address patriarchal gender relations as a route to empowerment. In both cases the cultural and religious components of women's oppression and gender hierarchies, as well as the potential of 'spiritual capital' (Iannaccone and Klick 2003) to transform gender relations and empower women, have been overlooked and under-theorized. While it is important not to reduce the subordination of women to religious or cultural causes, or to essentialize poor women as particularly religious or spiritual, the following discussion will indicate the relevance of religio-cultural factors to many of the concerns that GAD typically addresses.

Feminist transformations of religion can be an important element of a range of different processes that enable women to enhance their empowerment, where empowerment may be defined as 'women increasing their ability to act, to perceive themselves as capable, to hold opinions, to use time effectively, to control resources, to interact with others, to initiate activities, to respond to events' (Rowlands 1998: 23). The use of the term 'empowerment' by a range of different actors, from the World Bank to grassroots women's organizations in developing countries, and often for different purposes, has caused some to question the efficacy of the term. Concern that it has become a 'buzz word' (ibid.: 11) or a 'motherhood' term (Parpat 2002: 41) adopted by different users has led to the criticism that 'the appropriate meaning will be understood without being explained' (Rowlands 1998: 11). Early uses of the term were consistent within 'women in development' discourse (WID), which emphasized that empowerment was synonymous with bringing women into the development process in enabling them to participate and giving them a voice. This view, however, has been criticized for its failure to consider that access to participate in systems that are essentially biased against women is unlikely to be empowering unless the structures that discriminate against women are themselves challenged and dismantled. As Parpat writes, 'the romantic assumption that giving a voice to the poor, especially women, will solve gender inequalities is questionable' (2002: 54).

In line with a shift to a 'gender and development' approach, the use of the term 'empowerment' requires that the nature and structures of gender relations also be considered. In particular, the GAD view of empowerment 'has been strengthened by the theoretical advance which enabled a distinction to be drawn between women's practical and strategic gender needs' (Rowands 1998: 16). It is precisely within this GAD view of empowerment that consideration of religion becomes most relevant. The transformation of patriarchal religion in some contexts is one way in which women can enhance their strategic position within the gender hierarchy in their societies, thus confronting and overcoming the disempowering obstacles that restrict their development and empowerment. Moreover, in strongly religious contexts, particularly where the state and/or political parties are influenced by conservative religious attitudes and are wary of secular feminism as a western imposition, it can be expedient to approach gender issues within a religious framework.

As we have seen in earlier chapters, predictions about secularization and the eventual disappearance of religion as societies modernized are now widely discredited for being overly simplistic. Thus, any assumptions that the negative influence of religion on gender inequality would cease to be a problem, with increased economic and educational opportunities for both women and men (and therefore that there was little point in engaging with religion in the short term), are now being revised. Moreover, in recent years, increasing numbers of those working in gender and feminist studies – which are typically dismissive of religion – have begun to recognize the importance of religion in the lives of many women in developing contexts, and also to express doubts about the normativity of theories of secularization and secularism. Reilly (2011), for instance, points out that this new approach within gender/feminist studies is critical of the way in which secularism has influenced how the West deals with religion, particularly Islam, and that the identification of secularism with modernity and democracy – in the words of Judith Butler (2008: 2) – results in '"cultural assaults" on religious minorities' (2011: 15). Reilly concludes that an emancipatory feminism 'entails recognition of the complex and often contradictory intersectionality of women's identities and experiences cutting across gender, socio-economic privilege, ethnicity, religion, sexuality, geo-location, and so on' (ibid.: 26).

Explanations of the reasons for the contemporary global 'religious resurgence' are complex and debates are ongoing, but what they do indicate is that there is no straightforward relationship between economic development and a lessening of religiosity. This suggests the importance of building considerations of religion into the development process, not least with respect to considerations of the causes of and strategies to ameliorate gender inequality. However, this is not a straightforward exercise and in what follows we will unpack some of the challenges, as well as opportunities, facing the engagement of mainstream development with religion around women's rights. First, we will explore the ways in which attitudes towards women in religious traditions present challenges for mainstream development policy and practice; second, we will look at the engagement with 'religious feminisms' as a culturally located strategy for encouraging and realizing female empowerment; and finally, we will return to the concerns outlined above: that the 'turn to religion' by donor-driven development policy and practice has gender costs that have not been adequately considered.

Attitudes towards women in religions: challenges for GAD

The extent to which religious traditions teach that men and women are different and therefore afford them different human rights has been widely explored in the literature, particularly with respect to Islam (e.g. Rawi 2004; Helie 2004; Soares 2006; Jamal 2005; Moaddel 1998, 2002; Gerami and Lehnerer 2001; Afary 1997; Lachenmann 2004). Seguino's study, briefly discussed in the introduction, which demonstrates that no one religion stands out as being more gender-inequitable than another, would seem to undermine assumptions that Islam is particularly problematic for gender equality (therefore indicating that other factors are at play in this focus on Islam). Instead, in all religious traditions we can find examples of the tendency to treat women as different (and indeed often as secondary) to men, which not only has a bearing on their ability to hold positions of leadership within their churches, mosques or temples, but also has a direct impact upon their social status and roles in society more broadly. While in many Christian denominations women can now be ordained as priests, this is still forbidden in Catholicism and in the Orthodox tradition – and even in the more liberal Anglican denomination (where women have been able to be ordained since 1994) there is an ongoing debate over the ordination of women bishops. In Islam, women cannot lead 'mixed prayers' (i.e. involving men and women); in Hinduism the Brahmin priesthood is hereditary and is a position that can be held only by men; and while in Mahayana Buddhism (practised in China and Taiwan, for instance) women can fully ordain, in Tibetan and Theravada traditions (practised, for instance, in Thailand) there continue to be immense obstacles to women's ordination at the same level as men (Tomalin 2006b, 2009a; Plate 6.1). In all traditions, however, there are movements, led mostly by women, campaigning for women's access to ordination and other positions of leadership within their traditions. Often these movements are underpinned not only by a desire for women to be able to practise their religion at the same level as men, but also by recognition that religiously based gender inequalities have a broader impact on women's social status and opportunities within society.

Differences between men and women, often reduced to women's biological capacity to bear children, result in understandings of different roles and

Plate 6.1 *A Thai Buddhist nun (bhikkhuni) from the Theravada tradition, shortly following her ordination*

Source: Emma Tomalin.

responsibilities between the sexes and can account for the uneven treatment of men and women in religious traditions. Such biological determinism has been strongly rejected by feminists since the 1960s, yet is a feature of many versions of religious traditions, which consider that such differences are immutable and divinely sanctioned. Within religious discourses on gender it is common to find the statement that women *are equal to but different from* men, reflecting their God-given duty to bear children and look after their husbands. However, as we shall see, there are also various assumptions that women are less capable than men, or cannot be trusted to behave in accordance with religious moral frameworks. There can be no doubt that these religious attitudes towards women have had an impact upon their social status in society more broadly and that in addressing women's social disadvantage it is also important to examine and to attempt to transform religious attitudes. This is not an activity, however, that international development actors need to initiate, since there are already a wide range of individuals and organizations seeking to address women's rights from within a religious framework with which partnerships could be formed.

Nonetheless, the extent to which religious texts and teachings have a *direct bearing* on women's oppression has been debated, with some commentators urging caution in drawing such conclusions and using them as the basis for social action. We will explore this issue with respect to a Hindu text, the Manusmriti ('Laws of Manu'), a legal text written by Brahmins, the highest priestly class, between the second century BCE and the second century CE. Many view the Manusmriti as having played a central role in solidifying and institutionalizing negative attitudes towards women in India, evidenced by the observation that there was a gradual decline in women's fortunes from the early Vedic period (1700–1100 BCE) to that of the 'Laws of Manu'. Box 6.2 presents

---

Box 6.2

The Manusmriti

With respect to the 'proper' treatment of women, an excerpt from chapter 9 of the Manusmriti tells us that

3. Her father protects (her) in childhood, her husband protects (her) in youth, and her sons protect (her) in old age; a woman is never fit for independence.
4. Reprehensible is the father who gives not (his daughter in marriage) at the proper time; reprehensible is the husband who approaches not (his wife in due season), and reprehensible is the son who does not protect his mother after her husband has died.
5. Women must particularly be guarded against evil inclinations, however trifling (they may appear); for, if they are not guarded, they will bring sorrow on two families.
6. Considering that the highest duty of all castes, even weak husbands (must) strive to guard their wives.

Source: *Laws of Manu* (1886).

a passage from the text, which, while it might seem to affirm Basham's statement that 'the ancient Indian attitude to women was in fact ambivalent [, she] was at once a goddess and a slave, a saint and a strumpet' (1954: 182), certainly does not suggest that women were viewed as independent and trustworthy individuals. Instead, their primary role is as a wife and a mother.

However, whether or not such textual pronouncements are culpable for women's oppression in India, underpinning actual practices that oppress women and girls such as dowry violence, female infanticide related to son-preference, *sati* ('widow burning') and child marriage, has been much debated. Kishwar (no date), for instance, finds the linking of the Laws of Manu to women's oppression in so many accounts to be deeply problematic, objecting to the way in which this reading of the text makes it seem as though *all* Hindu women are oppressed, and for the same reasons. She argues that while the Laws of Manu were used by the British colonialists in the eighteenth century to develop Hindu personal law, this was a misappropriation and this one text was never authoritative in a pan-Indian manner; it only appeared to be so because the British liaised with the Brahmin class, who favoured this text. She writes that

> the disgraceful treatment of Dalits and downgrading of women are among the most shameful aspects of contemporary Indian society. But they will not disappear by burning ancient texts because none of the 'Hindu' scriptures have projected themselves as commandment-giving authorities demanding unconditional obedience from all those claiming to be Hindus.
>
> (ibid.)

The way to address this issue, she argues, is not through 'burning texts' but through an analysis of broader socio-economic dynamics and interrelationships.

Kishwar is not dismissing the impact of 'religion' per se on women's disadvantage, but is rejecting the idea that there is a codified textual tradition which impacts upon all women in the same way. To look at the impact of religion in this way is to overlook the localized ways in which religion shapes women's identities and social attitudes towards women in conjunction with other factors such as age, class or caste. It assumes a timeless, unchanging Hindu tradition, an attitude reflecting colonial orientalist attitudes towards religion. This analysis is also relevant to thinking about the link between religious traditions and gender inequality in other contexts and with respect to other religions. If we are to take seriously the role that religion plays in both women's oppression and their empowerment, then it is crucial that we do not oversimplify the role of religion as purely acting via its texts, or overemphasize its role and treat it in isolation to other factors. Nonetheless, religious attitudes about the nature of women and their roles in the family and the broader society can have a concrete and direct impact upon women's freedoms and how they are actually treated in certain contexts. In the following sections we will explore this with respect to two issues: religious fundamentalism and reproductive rights.

Religious fundamentalism

As I have already noted, within the literature on women's human rights and religion there has been an emphasis upon Islamic contexts, and this reflects the concerns of activists in both the Global North and the Global South about increasing Islamization, or Islamic fundamentalism, in some Muslim areas. In Nigeria, for instance, the adoption of Shari'ah law in the north of the country since 2000 has been shown to act to women's disadvantage. Kalu (2003), for instance, discusses well-publicized cases where women found guilty of adultery have been sentenced to death by stoning, which is viewed as an Islamic response to adultery. However, both local and international outcry has meant that few sentences have actually been carried out (Pearson and Tomalin 2008). Other studies focus on the links between Islam and domestic violence, including so-called honour killings, as well as the relationship between Islam and 'forced marriage' (Hajjar 2004; Sookhdeo 2008; Patel and Gadit 2008).

These practices are typically viewed as manifestations of religious fundamentalism, although, as is discussed in Chapter 4, such terminology is problematic, not least because the term is not normally employed by those it is used to describe and tends to be used by westerners in a negative sense, typically with reference to forms of Islam (Hawley and Proudfoot 1994). Moreover, attention has also been drawn to the way in which the plight of women within 'fundamentalist' contexts has been used to justify military campaigns by the West against Muslim countries and to the fact that such women 'figure in imperialist agendas in a thoroughly orientalist manner' (Stabile and Kumar 2005: 765; see also Ahmed 1992). Nonetheless, part of the so-called religious resurgence globally appears to be the rise of tendencies within all religions where we find the 'vigorous promotion and enforcement of gender roles whose explicit intent entails the subordination of women' (Howland 2001: xi). According to the founding statement of the international feminist organization Women Against Fundamentalisms (est. 1989), which has a long history of research and activism around gender issues:

> Fundamentalism appears in different and changing forms in religions throughout the world, sometimes as a state project, sometimes in opposition to the state. But at the heart of all fundamentalist agendas is the control of women's minds and bodies. All religious fundamentalists support the patriarchal family as a central agent of such control. They view women as embodying the morals and traditional values of the family and the whole community.
>
> (Brown *et al.* 1997: 80)

Thus, despite problems with the term 'fundamentalism', a number of scholars have argued that it is a useful concept pointing towards a modern phenomenon with shared characteristics across different religious traditions, including the control and symbolic idealization of women (Hawley and Proudfoot 1994: 27–30). Box 6.3 demonstrates the way in which gender is constructed in religious fundamentalism according to the analysis of Hawley and Proudfoot (ibid.: 27–35).

Box 6.3

The construction of gender in religious fundamentalism

The role of women as 'other' in fundamentalism

Women feature heavily in fundamentalist discourses for a number of reasons, not least because in the absence of being able to directly control the 'other' that exists externally to the fundamentalist group (e.g. the secular West), attention is redirected to women as a controllable 'other' within the fundamentalists' terrain. As Hawley and Proudfoot write:

> For every text that places well-domesticated womanhood on a religious pedestal, another one announces that, if uncontrolled, women are the root of all evil; and to the perception of many fundamentalists, the loosening of women is a prominent feature of modern Western secularism. Thus the focus of chaos is transferred from an external other to a familiar one.
>
> (1994: 27–28)

Woman as part of a nostalgia for an idealized past

> Since men primarily control the construction of this idealized past, their solution is to portray the women who inhabit it as self-sacrificing and generous.
>
> (Hawley and Proudfoot 1994: 30)

Women also play a role in symbolizing motherhood, which is an important symbol for fundamentalist groups that feel alienated from modern society (e.g. the importance of 'Mother India' to Hindu nationalism).

Religious machismo

According to Hawley and Proudfoot (1994: 27–35), men in fundamentalist movements need to reassert their masculinity in the light of threats. For them to do so successfully, the presence of women is necessary in order for them to be protected. These women may be symbolic or real (e.g. the importance of 'Mother India' to Hindu nationalism).

As Box 6.4 illustrates, religious fundamentalist groups such as the Taliban in Afghanistan hold conservative attitudes about the different roles of the genders. These are backed up by reference to 'tradition' – and, more importantly, to textual tradition: the texts are to be taken literally. Their goals are largely incompatible with those of western feminist and development agendas and are sometimes articulated in direct opposition to feminist ideas. Moreover, they frequently inflict harm or death upon women if they deviate from the path laid down for them. Over the past decades a number of grassroots women's organizations and coalitions have emerged to deal with the social implications of fundamentalist religion for women. For instance, Box 6.5 describes the work

Box 6.4

The Taliban in Afghanistan

According to the website of the Feminist Majority Foundation (which was founded in the United States in 1987 and has run its Campaign for Afghan Women and Girls since 1997), upon seizing power in 1994 the Taliban regime, which claims to follow a 'pure, fundamentalist Islamic ideology',

> instituted a system of gender apartheid effectively thrusting the women of Afghanistan into a state of virtual house arrest. Under Taliban rule women were stripped of all human rights – their work, visibility, opportunity for education, voice, healthcare, and mobility. When they took control in 1996, the Taliban initially imposed strict edicts that:

- Banished women from the work force
- Closed schools to girls and women and expelled women from universities
- Prohibited women from leaving their homes unless accompanied by a close male relative
- Ordered the publicly visible windows of women's houses painted black and forced women to wear the burqa (or chadari) – which completely shrouds the body, leaving only a small mesh-covered opening through which to see
- Prohibited women and girls from being examined by male physicians while at the same time prohibited female doctors and nurses from working.
- Women were brutally beaten, publicly flogged, and killed for violating Taliban decrees.

While the Taliban was defeated in 2001 and women have regained some of the rights that they were denied under the regime, 'many Afghan women and girls still live without basic necessities, nearly 1,000 girls' schools have been attacked since 2002'. The Taliban has also regained control of some areas of the country.

Source: http://feminist.org/afghan/taliban_women.asp (last accessed 11 September 2012).

of the organization Women Living Under Muslim Laws (WLUML), founded in 1984 in response to situations where women were being denied their rights by reference to Islamic law.

While secular legal instruments clearly play an important role in enabling women to protect themselves against the impact of fundamentalism, others have argued that these may not be sufficient, particularly in highly religious contexts and where there is suspicion of gender programmes (e.g. among male religious elites) for being a product of western, secular values. Instead, to approach women's rights from within different religious frameworks via engaging with 'religious feminisms' may prove more effective in some situations. Before examining the role of engaging with 'religious feminisms' as a route to women's rights and empowerment for GAD approaches, we will first look at the impact that religions have upon women's reproductive rights and choices.

Box 6.5

Women Living Under Muslim Laws

According to the organization's website:

> Women Living Under Muslim Laws is an international solidarity network that provides information, support and a collective space for women whose lives are shaped, conditioned or governed by laws and customs said to derive from Islam.

> For more than two decades WLUML has linked individual women and organisations. It now extends to more than 70 countries ranging from South Africa to Uzbekistan, Senegal to Indonesia and Brazil to France. . . .

It aims to challenge

> the myth of one, homogenous [*sic*] 'Muslim world'. This deliberately created myth fails to reflect that: a) laws said to be Muslim vary from one context to another and, b) the laws that determine our lives are from diverse sources: religious, customary, colonial and secular. We are governed simultaneously by many different laws: laws recognised by the state (codified and uncodified) and informal laws such as customary practices which vary according to the cultural, social and political context.

Source: www.wluml.org/node/5408 (last accessed 11 September 2012).

Reproductive rights and choices

Women's rights activists globally have expressed concern about the ways in which 'conservative' religious traditions influence women's access to reproductive services such as contraception and abortion, therefore denying them their reproductive rights (Amin and Hossain 1995; El Dawla 2000; Kazimov 2003; Obermeyer 1994; Bowen 1997; Hasna 2003; Borland 2004). In addition to this, religions can also have a powerful influence on the reproductive choices that women make, which in turn may result in threats to their reproductive rights. Razavi writes that this

> has been evident in the alliance forged between some Islamist states and the Vatican (in the context of the United Nations conferences of the 1990s) in opposition to the demands of global women's movements for gender equality, and most explicitly in reproductive and sexual rights.
>
> (2009: v)

Since the 1994 United Nations Population and Development Conference in Cairo, the idea of reproductive rights has increasingly dominated discussions about issues around family planning (Grimes 1998; Smyth 1998). There was a shift to 'a new and broader reproductive rights approach that is fundamentally different from the prevailing family planning and demographic approach to population' (Bandarage 1998: 7). However, Bayes and Tohidi (2001) observe the emergence of a strategic Muslim–Catholic alliance, which was particularly visible at the 1995 Beijing women's conference and continues to be suspicious of current approaches to reproductive rights. An analysis of religious speeches that were made during the 1995 Beijing conference was undertaken by Couture (2003), and she expresses concern about the way in which a number of participants argued that women's reproductive choices should be defined by religious values. This points to a tension within the international women's movement (represented at the Beijing conference by both secular and religious actors) between universal rights and cultural or religious particularisms (Vuola 2002).

Powerful North American voices have also been instrumental within this conservative religious backlash against reproductive rights through the 'Mexico City Policy' (more recently reincarnated as the so-called 'global gag rule'). The Evangelical religious right in North America was influential in the former president George W. Bush's reinstatement of the 'Mexico City Policy' on his first day in office in January 2001 (Crane and Dusenberry 2004). First introduced by the Reagan administration in 1984, it was overturned by the Clinton administration in 1993 and again repealed by President Obama in 2009. It prohibited any organization – domestically and internationally – in receipt of US funds from providing abortions (except in the cases of pregnancy from rape or incest, or when a woman's life was in danger), and even from providing advice and information about abortion services, or engaging in any lobbying activity for the legalization of abortion. This policy has had a detrimental effect on the provision of reproductive and health services for women that goes far beyond the apparent curtailment of abortion-related services (Rose 2005: 1211; Cohen 2001; Centre for Reproductive Rights 2003).

However, the impact of these restrictive attitudes towards contraception and abortion has been most strongly felt in the staunchly Catholic Latin American nations. Kane (2008), for instance, explores the influence of the Catholic Church on recent changes in abortion policy in Nicaragua, Colombia and Mexico City (see also Sánchez Fuentes *et al.* 2008). In all three contexts, women's activism for abortion rights intensified in the wake of the UN women's conferences of the 1990s. However, this activism took place against the backdrop of opposition from the Catholic Church, with support from growing numbers of Evangelical Protestants. While in Colombia there was modest success, resulting in legislation in 2006 to permit 'therapeutic abortions', and also in Mexico City, where in 2007 all abortion in the first trimester became legal, the story was very different in Nicaragua. Here an alliance between Catholics and Evangelical Protestants was influential in achieving an outright ban on abortion in 2006, resulting from a change of heart by the former Sandinista president, Daniel Ortega. During his period in office in the 1980s he had supported abortion, but during his 2006 election campaign he opposed it in order to gain Catholic votes. Thus, in some contexts religious opposition can be effectively met by the use of international (secular) law to support arguments about reproductive health and rights (as in Colombia and Mexico

City). However, the need to be vigilant and to protect even limited access to abortion is underscored by the case of Nicaragua. Political considerations may often trump support for women's rights in contexts where the Church is powerful.

The impact of Catholic restrictions on birth control has also had far-reaching impact upon high maternal mortality rates in many Catholic contexts. A recent article in *Time* magazine, reporting on Catholic attitudes towards birth control in the light of the co-hosting of the London Summit on Family Planning (11 July 2012) by DFID and the Bill and Melinda Gates Foundation (Melinda Gates is herself a practising Catholic, although not of the conservative variety), tells us that:

> A new Johns Hopkins University study, financed by the Gates Foundation and published this week in the British science journal the *Lancet*, shows that increased contraceptive use could cut maternal mortality in developing countries by a third, not just by lowering unhealthily excessive childbirth rates but also by helping to avoid risky teen pregnancies and reducing unsafe abortions. . . .
>
> In developed regions like Western Europe, maternal mortality rates can be as low as five female deaths per 100,000 live births, as it [*sic*] is in Sweden. But in Nicaragua, where the Catholic Church can still bully the government into making contraception scarce, it leaps 20-fold to a frightful 100 deaths per 100,000 births – and teens account for 45% of the country's pregnancies, one of the highest levels in the Americas. Contrast that with neighboring Costa Rica, where birth control codes aren't legislated from a bishop's pulpit, and the rate is less than half of Nicaragua's, at 44 deaths per 100,000 births. The situation is vastly worse in Africa, where maternal mortality is 430 per 100,000 births in Uganda, whose Catholic Church is politically powerful, and it reaches almost triple that in some other countries on the continent.
>
> (Padgett 2012)

However, not all Catholics are opposed to birth control and even abortion. For example, Catholics for Choice (CFC), founded in 1973 in the United States, is engaged in both pro-choice and pro-condom campaigns and argues that this stance is compatible with Catholic religious teachings.

In addition to restricting women's reproductive rights through influencing the availability of reproductive services, religious traditions may also shape religious values that have effects on the *reproductive choices* that women make. In terms of reproductive choice, the issue of 'son preference' and the 'reversed sex ratio' in India and China have received much critical attention. In both countries we find 'reversed sex-ratios', where the sex ratio of the overall population is 'masculine' (there are more men than women), whereas in most other parts of the globe it is 'feminine' (Sen 1990: 61). While these differences can be at least partially explained in terms of the level of care that females receive within the particular societies (i.e. in terms of diet and access to health services), they have also been linked to female infanticide and the rise in sex-selective abortions (Sen 1993, 2003; Klasen and Wink 2003; Klasen 1994).

Female infanticide and the sex-selective abortion of female foetuses, particularly in India and China, have been given an economic explanation. In China the one child policy, in force since 1979, has given rise to a marked preference for boys, in a culture where males are more economically productive and are expected to care for parents in their old age. In India the practice of female infanticide and sex-selective abortion of female foetuses has been linked to the spread and intensification of the custom of dowry, where money and/or gifts are given by the bride's family to the groom's family at marriage, which similarly supports a preference for male children (Basu 1999; Banerjee 2002; Plate 6.2). What was once a practice largely confined to upper-caste Hindus in the north of India has now spread to other castes, religions and regions of South Asia (Rozario 2009). The reasons for this are complex and are, at least partially, dependent upon a combination of 'Sanskritization' (upper caste emulation), neo-liberalization and rise of cash incomes (Banerjee and Jain 2001; Banerjee 2002). There is also some evidence that the size of dowries has grown in recent decades, putting increasing pressure on families to prefer sons to daughters (Bradley *et al.* 2009; Basu 1999; Banerjee 2002; Dreze and Sen 2002; Jayaraj and Subramanian 2004; Sen 1992, 1993; Sunder Rajan 2003). The pressure on females in respect of dowry, however, not only begins before birth but extends after the time of marriage. Demands for dowry often continue after a bride is

Plate 6.2 *Government-sponsored public sign campaigning against female infanticide, Kolkata, India, 10 September 2007*

Source: Rita Banerji/Getty (10 September 2007).

married and can result in violence towards, or even murder of, women if they are unable to meet them.

While the practices of sex-selective abortion and female infanticide have been given economic explanations and legal solutions have been set in place (though these have not managed to ameliorate the problem), cultural or religious factors that advocate a preference for sons should also be considered. Neither the one-child policy nor the giving of dowries is a religious practice, yet the extent to which economic explanations dovetail with religious and cultural understandings about the role of women has been noted. In China, for instance, the strong Confucian preference for boys would seem to add to female disadvantage when coupled with the one-child policy. The Confucian tradition places a strong emphasis upon sons above daughters, for both cultural and economic reasons: sons are necessary to perform rituals to the ancestors, and there is a cultural and economic expectation that sons will care for their parents in old age. Hence, the one-child policy has had a negative impact upon the female population (Junhong 2001; Feng 1996; Greenhalgh 2001; Li 1995), since families would rather have one son than one daughter.

In India, where notions of female roles and responsibilities are ingrained in and supported by many people's religious beliefs and practices, the interfaces between religion, gender and dowry need further investigation, as they are important in understanding the interconnecting factors that sustain abusive dowry practices. This could help us understand the cultural underpinnings of the dowry problematic, as well as influence forms of activism to curb dowry. In particular, it is important to understand the ways in which religion helps shape and sustain gender hierarchies in India but also its potential to challenge patriarchy.

What this discussion suggests is that to oppose the religious obstacles to women's rights through the use of secular legal strategies may not be appropriate in every situation. Instead, in the next section we explore the argument that to engage with 'feminist' or gender-equitable interpretations of religious texts and traditions, by supporting influential individuals or organizations that undertake this approach, might be a more effective way of transforming attitudes and diminishing gender inequality. For instance, dowry has been illegal in India since 1961, but continues to spread and intensify. While this suggests that the legal system is not performing as well as it should, it is also the case that dowry practices both rely upon and reinforce the inferior status of women, and a better understanding of the role that religion plays in this process could make a useful contribution (Bradley and Tomalin 2010). This further reinforces the important point that religion in societies interacts with other social forces. It is part of a nexus of factors that shape people's opinions and values, including those relating to gender. To reduce social phenomena to religious causes is likely to limit our analysis, yet to ignore the role that religion plays alongside economic or political influences is also limiting.

Engaging with 'religious feminisms'

While some studies are critical of the ways in which religion can be used to restrict women's rights, and advocate a secular approach to rights instead, others emphasize the potential benefits of finding correlates for rights within religious traditions (Engineer 2004; Ali 2000; Kardam 2005; Afkhami and Friedl 1977; Afkhami 1995; Sechzer 2004; Donno and Russet 2004; Nussbaum 2000; Drumbl 2004). Thus, religion is seen as a barrier to women's human rights but also as an important vehicle for changing attitudes that obstruct the realization of rights. In this 'religious feminist' discourse we frequently find the argument that religious traditions were more gender-equal when they were first formed or in their earlier periods than they later become under the influence of patriarchy (Cooey *et al.* 1991). However, the challenge for 'religious feminists' today is how to disentangle patriarchal values from their justification as religious, and to promote alternative (many would argue 'authentic') interpretations of religious traditions that are supportive of women's human rights and empowerment. 'Feminist' organizations, such as Sisters in Islam in Malaysia, are undertaking such interpretive work as well as working at the grassroots level to transmit more egalitarian versions of Islam to women (Foley 2004; Sleboda 2001).

Thus, while religion clearly acts against women's interests in many contexts, particularly when its control is in the hands of a conservative male elite, to engage with 'religious feminisms' may in certain circumstances be more effective than to employ secular strategies. Adamu, for instance, writing about Nigeria, argues that 'because gender issues are both religious and political concerns in many Muslim societies . . . any attempt to reform gender relations that excludes religion is likely to fail' (1999: 56). She suggests that because international donors have not incorporated Islam into their gender policies and programmes, there is a perception that GAD is a tool of the West and that some of the principles adopted in international documents, such as CEDAW (the United Nations Convention on the Elimination of all forms of Discrimination Against Women), are seen to be at odds with Islamic values about the family. While there are Muslim feminists in many Islamic contexts who are committed to the pursuit of women's rights within an Islamic framework, they tend to find themselves trapped between conservatives in their own tradition and critics from the West who believe that in using Islam as a framework from which to approach rights, they are ultimately 'accepting or supporting their own subordination' (ibid.: 57). Adamu points out that by the 1990s, donors in the Global North were beginning to fund Islamic NGOs working on women's rights, but that relationships between the different actors remained tense.

One of the challenges facing this approach is the fact that visions of what counts as 'female empowerment' may differ, since 'religious feminisms' do not necessarily map neatly onto GAD goals. Arguably, there is a need for a better understanding of what constitutes notions of women's empowerment in different contexts, and for an openness to include a diversity of views and strategies even when they might seem to conflict with secular feminisms premised on female and male equality in all spheres of life. Deneulin and Banu draw a contrast between secular women's rights NGOs in Pakistan, which tend to stress 'individual liberty, including sexual liberty, and participation of women in economic and political affairs' (2009: 160), and female madrasas, for which the focus is on women's interests being 'best served in a stable family unit . . . [where] the emphasis is not on equality but on equity' (ibid.: 160). They argue that NGOs in the Global North have tended to engage with the secular westernized elite feminists in Pakistan and have marginalized the female 'Islamic leadership' (ibid.: 285) in the madrasas. These different understandings of empowerment suggest the need for 'dialogue', where each group attempts to understand the other's point of view. Moreover, research that addresses both tensions and successes in interactions between secular and religious actors in relation to gender concerns could be useful in highlighting areas where there is likely to be agreement, as well as potential flashpoints that need to be carefully negotiated, or even avoided.

One of the implications of this discussion about engaging with 'religious feminisms' is, as Razavi and Jenichen point out, that the line between different feminisms, secular and religious, is more blurred than we once thought, where 'those who work for "internal reform" [of religious traditions] very often draw on the ideas and arguments of "external" advocates for change. Alliances between feminists of different religious and secular communities are therefore imperative' (2010: 846). These alliances can be seen, for instance, in the strengthening of transnational women's movements, constituted by networks of regional groups and movements that promote their own strategies for women's empowerment. The aim of using the term 'transnational' is to describe an observed shift in feminist styles of organization where, although 'feminist groups and women's organizations remain rooted in national or local issues . . . their vocabulary, strategies and objectives have much in common with each other and have taken on an increasingly supra-national form' (Moghadam 2000: 62). Organizations such as Women Living Under Muslim Laws and Catholics for Choice, as well as movements within religious traditions concerned with issues such as female ordination, all suggest the growth of vibrant transnational feminist networks that aim to negotiate women's empowerment from local religious and cultural contexts rather than from earlier styles of international feminism, which were often perceived as western-driven and top-down, favouring 'secular' approaches and values (Tomalin 2009a; Moghadam 2000).

Many have welcomed the fact that, over the past decade or so, mainstream development actors have started showing a far greater interest in funding and working with FBOs and religious leaders, including engaging with various 'religious feminisms'. As we shall discuss in the next section, others, however, are more cautious because the 'turn to religion' may have negative gender-related consequences that have yet to be adequately considered by development practitioners and policy makers. It is also the case this recent 'turn to religion' has meant that there is often an assumption that 'a feminist re-engagement with religious texts within a religious framework is a panacea for altering gender bias in laws, policies and practices' (Tadros 2011: 9). Bartelink and Buitelaar take up this critique in their analysis of a project run by the Dutch government in Yemen that aimed to understand how women 'draw on religious and cultural resources to claim rights to reproductive health and education' (2006: 352) and to incorporate Muslim women's experiences into policy. They conclude that the project set out to prove the assumption that Islam could empower women rather than to test it. For instance, they found that women were reluctant to use feminist interpretations of the Qur'an when presented with them – perhaps as a reaction against perceived western interference – and that they stuck to 'traditional' interpretations as often as men did.

The rush to 'find the religious': engaging with religious leaders and 'faith-based organizations'

Very little research has been carried out to date on the gender-related implications of current development policies and practical initiatives that actively engage with religion (those that explicitly engage with gender as well as those that do not). Balchin gives an example:

> I recall the fury of a women's rights activist from the Mindanao region in the Philippines in the 1990s. A foreign bilateral agency had apparently gathered local ulema [male Islamic scholars] in order to produce a statement supporting women's reproductive rights from an Islamic perspective. Although this was not a normal sphere of the local ulema's concern, the statement was duly issued, but more importantly, the gathering facilitated networking among the ulema that subsequently contributed to the formation of a political grouping that promoted a fundamentalist vision of Islam. In other words, a development approach reinforced conservative interpretations of religion and strengthened the power of those who do not have pluralism and equality at heart.
>
> (2011: 17)

As Balchin warns, 'this rush to "find the religious" is rarely backed by sophisticated knowledge of the diversities among religious groups' (ibid.: 17). Moreover, there is often an uncritical adoption of dominant (usually male) perspectives and voices within religious traditions as though they are representative of the tradition as a whole. This runs the risk of marginalizing other voices and positions that may not have such a prominent public presence – specifically, feminist or gender-sensitive interpretations within religious traditions (Tomalin 2009a). Considering the highly patriarchal nature of most religions, women's participation in religious institutions (e.g. churches and mosques) and 'faith-based organizations' is likely to be marginal. Even when women do have opportunities for leadership in FBOs, such leadership does not necessarily challenge traditional gender regimes. Moreover, access to services or assistance may also be conditional 'on conforming to the FBO's interpretation of religiously appropriate gender roles and behaviour' (Tadros 2010: iii). Thus, while religion can provide women with coping strategies and concrete support services, these may also involve gender costs.

This 'turn to religion' is part of a broader 'global faith agenda' that has risen in significance at least since 9/11, but can also be guilty of promoting religion at the expense of other identities and approaches (Patel 2011). FBOs and religious leaders are often typified as being closer to the grassroots and therefore in a position to reflect women's best interests, as well as to provide support for their 'spiritual development' alongside their 'material development'. This may be true of some, but the ways in which others 'seek to impose their own values and ideologies rather than respond to so-called local gender agendas' (Tadros 2010: 11) should not be overlooked.

Moreover, the fact that a certain FBO might challenge problematic gender relations in some issues yet reinforce them in other areas is a further consideration for development practitioners. Tadros describes a situation in Malawi where Muslim, Catholic and Protestant FBOs opposed harmful widowhood practices, while at the same time opposing a government campaign promoting condom use (2010: 11). This points to the importance of grounding engagement with religion (in terms of organizations or leaders) in empirical knowledge of the location, and not ignoring the gender politics of particular religious institutions or FBOs and the broader social, political and economic context in which they are embedded.

However, this may be easier said than done. One of the challenges facing development practitioners is the difficulty of finding out about the nature of an FBO's approach to gender issues. Moreover, they may be wary about openly critiquing religious organizations or leaders if their attitudes towards gender are seen as problematic. Conversely, staff sometimes feel reluctant to engage with religion and religious organizations, citing ignorance as an excuse for 'not wanting to interfere in local culture' (Hopkins and Patel 2006: 434). These problems point to the need to develop understanding and trust with faith communities and to engage in activities that are guided by in-depth knowledge of the particular situation. It is also important for development practitioners to enlarge their repertoire of religious knowledge and to promote 'religious literacy' as an important skill for those involved in development. However, in contexts where secular organizations find it difficult to gather information about the gender attitudes in particular religions or to critique them when they are found to be problematic, one solution is for other FBOs to undertake this work. For instance, Marshall and Taylor (2006) discuss the work of the UK-based Christian NGO Tearfund, which works through local Evangelical Christian partner organizations in over seventy countries. A project was set up for the purpose of establishing the extent to which Evangelical churches in Africa were dealing with issues around sexual activity and behaviour, and whether they were doing so in an informed and progressive manner. One of the findings of the project was that where churches taught that women should serve and be subservient to men, women were less likely to be able to 'challenge the unfaithfulness of their husbands, or negotiate the use of a condom for safer sex' (ibid.: 370). Overall, Tearfund was critical of the Evangelical churches – a position that is perhaps easier for it to adopt as an 'insider' to the tradition – and was funding a small pilot initiative aimed at changing attitudes.

Summing up

The aim of this chapter has been to address the following question: *Is religion bad for women's development and the pursuit of gender equality?* We examined the emergence of a GAD approach within mainstream development policy and practice, but the chapter highlighted the ways in which this has tended to avoid or marginalize considerations of religion. It has done so despite strong evidence not only that religion influences gender hierarchies but also that engaging with 'religious feminisms' can help transform religious traditions in contexts where they are complicit in compounding women's disadvantage. However, the marginalization of religion has begun to recede, following the recent 'turn to religion' by donor-driven development policy and practice. A wider range of GAD-related programmes seeking engagement with religious leaders and movements have been developed. Nonetheless, despite the potential benefits of this 'turn', there is some concern that much engagement with religious leaders and FBOs has gender costs that have not been adequately considered. Neither, as Sholkamy warns, should we assume that 'religious feminisms' are the best or only way of pursuing women's empowerment, and we need avoid this 'pathway' becoming 'more important than the destination' (2011: 49).

A number of commentators have drawn attention to the tendency for engagement with religion in pursuit of women's rights to view poor women as particularly religious or to consider the difficulties that they face to be wholly reducible to religious or cultural causes (e.g. Bartelink and Buitelaar 2006).

The importance of avoiding these essentialist positions is revealed in an article by Hopkins and Patel (2006), which discusses two workshops held by Oxfam in 2004 and 2006. These workshops aimed to encourage Oxfam staff to explore their work on gender issues in 'Muslim contexts', and to 'understand the impact of religious, political, and other identities on gender equality' (ibid.: 424). The participants, who were mostly Muslim, promoted the view that religion should be considered, since there was a need to respond to local knowledge when approaching women's rights. However, they also argued that it was crucial not to prioritize religious identities over other identity markers or to assume that Muslim women were particularly religious. Moreover, they stressed that when talking about 'Muslim contexts', 'we should bear in mind that contexts are constantly evolving, and that our understanding of context is both time- and place-specific, and subjective' (ibid.: 426). Comparisons between the 2004 and the 2006 workshops demonstrate how quickly the terrain can shift, with participants highlighting the ways in which the increasing politicization of Islam *and* western military action placed further pressure on many Muslim women. The authors concluded that strategies for engagement need to be ever more sensitive, cautious and contextual, and that 'key allies' (e.g. religious leaders or 'faith-based' groups) must be carefully identified to ensure that they are committed to diversity and gender equality. Thus, while engagement with religion does offer potentially fruitful partnerships that could enable better development outcomes for both women and men, many development actors seem in rather a hurry to engage with religious leaders and 'faith-based organizations', thereby running the risk of colluding with dangerous forms of essentialism about poor women and cutting off the possibility of pursuing potentially more useful 'secular' strategies.

However, concerns about the tendency to essentialize and romanticize the role of religion within development discourses are not only relevant to a consideration of gender and women's empowerment, but also important to other aspects of international development theory and practice. In the next chapter, we will critically examine the view that this tendency is also prominent within discourses about religions and environmental sustainability.

Questions for discussion

1 How has religion been typically viewed within 'gender and development' (GAD) approaches?

2 In what ways have religious traditions had a negative impact on the definition and the realization of women's rights?

3 What is 'religious feminism' and to what extent could it play a role in women's empowerment?

4 What is meant by the 'rush to find the religious' and in what ways has this been problematic for women?

Recommended reading

IDS Bulletin **special issue (2011): Gender, Rights and Religion at the Crossroads (42 (1)).** This publication argues that this is a crucial moment for engaging with the politics of religion and gender. Since the events of 9/11, 'Muslim communities' have emerged as political categories, and this emergence has specific implications for thinking about and addressing women's human rights.

Journal of International Women's Studies **special issue (2009): Gender and Islam in Asia (11 (1)).** The articles in this special issue focus on gender and Islam in Asia in a bid to address a lack of publishing from and about this region when compared to the Middle East. Many of the articles are concerned with addressing the ways in which women work within an Islamic framework to pursue empowerment.

Third World Quarterly **special issue (2010): The Unhappy Marriage of Religion and Politics: Problems and Pitfalls for Gender Equality (31 (6)).** This special issue draws upon working papers from an UNRISD (United Nations Research Institute for Social Development) research programme, 'Religion, Politics and Gender'.[1] It explores religion as a political force and the ways in which it influences the struggles for gender equality in different contexts.

Tomalin, Emma (ed.) (2011) *Gender, Faith and Development*, **Rugby: Practical Action Publishing and Oxfam GB.** This edited volume comprises a selection of papers from special editions of the journal *Gender and Development* that focus on the relationships between religion, gender and development (issue 7 (1), March 1999: Gender, Religion, and Spirituality; issue 14 (3), November 2006: Working with Faith-Based Organizations; and issue 16 (2), July 2008: Reproductive Rights: Current Challenges). Downloadable at http://policy-practice.oxfam.org.uk/publications/gender-faith-and-development-144042.

Women's Studies International Forum **special issue (2010): From Village Religion to Global Networks: Women, Religious Nationalism and Sustainability in South and Southeast Asia (33 (4)).** The articles in this special issue come from an international workshop held in New Delhi in 2007. The workshop explored the influence of the increase of politicized religion on aspects of 'traditional' religious behaviour that can be seen as positive and supportive of human and environmental sustainability, as well as its impact for inter-religious and community relations as communal barriers intensify.

Websites

www.wluml.org 'Women Living Under Muslim Laws is an international solidarity network that provides information, support and a collective space for women whose lives are shaped, conditioned or governed by laws and customs said to derive from Islam.'

www.womenagainstfundamentalism.org.uk 'Women Against Fundamentalism (WAF) was formed in 1989 to challenge the rise of fundamentalism in all religions.'

Both websites have information about the work of the respective organization as well as downloadable publications.

7 Environmentalism, religions and development

This chapter will:

- address the following questions: *Is it correct to assume that religious traditions teach that one should care for the natural environment? Is this relevant in practice? In what ways might religion contribute towards 'sustainable development'?*;
- discuss the ways in which religious traditions are considered to teach that one should care for the natural environment;
- examine the origins and nature of *religious environmentalist* discourses;
- explore concerns about the impact of romanticized *religious environmentalist* discourses upon mainstream development policy and practice concerned with 'sustainable development', through looking at the impact of ecofeminism upon the 'women environment development' (WED) approach.
- examine the beneficial role that religion might play in 'sustainable development' initiatives within international development organizations, with a particular focus upon biodiversity conservation.

Introduction

Voices from religious traditions on the topic of humanity's relationship to the natural world have been evolving since the 1960s alongside the birth of the modern environmental movement. Broadly, they argue that religions consider nature to be significant beyond its use value to humans (albeit in different ways, according to particular religious teachings) and that people ought not to act in ways that harm the natural environment. Thus, religious traditions are considered to provide frameworks for environmental ethics and to support the view that nature should be treated with respect. However, we also find the argument that notions of the sacredness of the natural world have become lost in the modern era, with the emergence of capitalism, industrialization and secularization. Within this *religious environmentalist* discourse, which is found within all religious traditions, humanity is considered to have largely 'forgotten' the sacredness of nature, and this, it is argued, needs to be rediscovered in order to address the contemporary global environmental crisis. In this way, the range of ecological problems facing humanity, in both the Global North and the Global South, are given an ethical significance: environmental destruction is caused by humans and the only way to counter such destruction is to 'revive' religious attitudes towards nature that place limits on our interactions with the natural world.

While much religious environmentalist literature stresses a positive correlation between religious injunctions to care for nature and environmentally friendly behaviour, there are other studies that suggest a more cautious approach. These studies argue instead that the links between religions and the environment tend to be romanticized and in practice we should not assume that poor people who practise, for instance, Eastern religious traditions or indigenous religions (which, it is suggested, are more amenable to ecological interpretations) are inherent environmentalists (Tomalin 2009c). Pedersen, for instance, argues that claims about the environmental nature of religion are 'anachronistic projections of modern phenomena onto the screen of tradition' (1995: 264). This chapter addresses the following questions: *Is it correct to assume that religious traditions teach that one should care for the natural environment? Is this relevant in practice? In what ways might religion contribute towards 'sustainable development'?* We will begin with an examination of where these religious environmentalist discourses are generated and by whom. Do they reflect romantic western notions of a lost ecological idyll (an ecological 'golden age') or do they have relevance for the poor who are struggling against floods, famines and droughts? When we find these discourses in developing contexts, are they more likely to be adopted by an educated 'middle class' that has been influenced by western 'eco-centric' and romantic approaches to environmentalism and sustainable development, or do they have a relevance at the grassroots as well?

Because of a range of factors, including poverty, a reliance on the land and geographical location, many of the poor in developing countries are particularly vulnerable to the consequences of environmental problems, and therefore, as outlined in the seventh Millennium Development Goal, presented in Box 7.1, 'sustainable development' needs to be a crucial dimension of the development process. In the final sections of the chapter we will consider the extent to which religion has a role to play in achieving 'sustainable development' in developing contexts where levels of religiosity are often high. Following MDG 7, what role might religion play in biodiversity conservation, the lobbying of governments and industry to adopt more sustainable policies and practices, and appealing to individuals to change their behaviour where it is contributing to environmental problems, in order to positively contribute towards improving people's livelihoods and well-being? Could religion in developing countries have a positive impact on reversing climate change and preserving species, in the interests of people's long-term development, or is this a 'post-materialist' romantic myth that essentializes poor people as inherently environmentalist?

The genesis of religious environmentalism

Religious environmentalist literature began to emerge alongside the modern environmental movement during the 1960s. Religious teachings have been reinterpreted, extended or synthesized in order to express concern for the environment, old rituals have been given new content and new rituals have been

---

Box 7.1

Millennium Development Goal 7: 'ensure environmental sustainability'

Target 7a: Integrate the principles of sustainable development into country policies and programmes; reverse loss of environmental resources

Target 7b: Reduce biodiversity loss, achieving, by 2010, a significant reduction in the rate of loss

Target 7a and 7b Indicators

7.1 Proportion of land area covered by forest
7.2 CO_2 emissions, total, per capita and per \$1 GDP (PPP)
7.3 Consumption of ozone-depleting substances
7.4 Proportion of fish stocks within safe biological limits
7.5 Proportion of total water resources used
7.6 Proportion of terrestrial and marine areas protected
7.7 Proportion of species threatened with extinction

Target 7c: Reduce by half the proportion of people without sustainable access to safe drinking water and basic sanitation

Target 7c Indicators

7.8 Proportion of population using an improved drinking water source
7.9 Proportion of population using an improved sanitation facility

Target 7d: Achieve significant improvement in lives of at least 100 million slum dwellers, by 2020

Target 7d Indicator

7.10 Proportion of urban population living in slums.

Source: reproduced from www.undp.org/mdg/goal7.shtml (last accessed 11 September 2012).

---

devised to reinforce the idea that there is a need for a re-evaluation of humanity's relationship to the natural world (Gottlieb 1996; Apffel-Marglin and Mishra 1993; Apffel-Marglin and Parjuli 2000; Bopp 1985; Chapple 1995, 1998; White 1967). Considering the increasing seriousness with which ecological concerns have been taken since the 1960s in both the Global North and the Global South, it is not surprising that this topic has warranted attention from within religious traditions, but also that activists and academics have expressed an interest. The emergence of so-called *ecotheology* has been an important development across all religious traditions and currently there are two peer-reviewed journals dedicated to this area: *Worldviews: Environment, Culture, Religion*, published by Brill; and the *International Journal of Religion, Nature and Culture*, published by Equinox.

Many consider that the modern debate about the relationships between religions and the environment began with Lynn White Jr's seminal 1967 article 'The Historical Roots of Our Ecological Crisis'. In it he argued that religion was the cause of, as well as the potential solution to, environmental problems. He wrote that 'Christianity bears a huge burden of guilt' for the environmental crisis facing humanity (ibid.: 1206) and considered that the displacement of paganism by Christianity had led to the desacralization of nature. The Christian God was seen as transcendent to creation, and humanity was granted dominion over the natural world. This, he argued, opened the way for humans to begin the domestication and transformation of nature. On the one hand, the growth of science was predicated upon the desire to understand God's 'mind', and on the other hand the development of technologies to manipulate the natural world were the 'realization of the Christian dogma of man's [*sic*] transcendence of, and rightful mastery over, nature' (ibid.: 1206). In tracing the roots of the contemporary environmental crisis to the union of science and technology in the nineteenth century, which created the conditions for the Industrial Revolution, he concludes that 'more science and more technology are not going to get us out of the present ecological crisis until we find a new religion, or rethink our old one' (ibid.: 1206). He suggests that religious traditions such as Zen Buddhism hold an environmentally friendly view of the humanity–nature relationship: such eastern traditions are represented as emphasizing the interconnectedness of humanity and nature, rather than their separation (see also Callicot and Aimes 1991). Or that a stewardship model, which draws upon the example of St Francis of Assisi, subverts and challenges the dominant attitudes towards nature, which are 'deeply grounded in Christian dogma' (White 1967: 1207).

Lynn White Jr was writing at a time when people were first becoming aware of, and were trying to make sense of, the nature and consequences of the environmental damage that humanity had been increasingly inflicting on the natural world, particularly since the Industrial Revolution. It was the publication of Rachel Carson's famous book *Silent Spring* in 1962 that brought to the attention of the American public the devastating effects of pesticides upon the countryside and the food chain. Paul Ehrlich (1968) and Barry Commoner (1971), respectively, drew attention to concerns over population growth and the inability of the Earth to deal with the effects of mass technologies. In 1970 the Club of Rome, a group of politicians, industrialists and scientists, developed a computer simulation model to predict the effects of environmental problems such as population growth and industrial expansion. Their report, *The Limits to Growth* (Meadows *et al.* 1972), predicted, for example, that famines and resource depletion would pose serious threats to life on the planet by the end of the century. In 1972 the first United Nations international environment conference was held in Stockholm and was attended by the Indian prime minister, Indira Gandhi, among others. This global response also gave rise to the production of a number of reports during the 1980s that helped to publicize the idea of a global environmentalism, including the Brandt Report (Brandt 1980) and the Brundtland Report (WCED 1987).

By 1983 the United Nations had set up a commission – the Brundtland Commission – to address concern 'about the accelerating deterioration of the human environment and natural resources and the consequences of that deterioration for economic and social development' (United Nations 1987a). This recognized the global nature of environmental problems and that it was therefore in the interest of all nations to work together and establish solutions to achieve 'sustainable development'. The report of the Brundtland Commission, *Our Common Future* (WCED 1987), defined sustainable development as 'development that meets the needs of the present without compromising the ability of future generations to meet their own needs' (United Nations 1987b) and 'called for international strategies combining both environmental protection and development' (Taylor *et al.* 2005: 1680). At the root of mainstream thinking about sustainable development is the idea that three dimensions need to be in balance – environmental, social and economic sustainability (Adams 2006) – and that one cannot be pursued in the absence of thinking about the others. For example, as Adams writes, the IUCN (International Union for Conservation of Nature) Programme 2005–8, adopted in 2005, used a model based upon interlocking circles 'to demonstrate that the three objectives need to be better integrated, with action to redress the balance between dimensions of sustainability' (ibid.: 2). This model is depicted in Figure 7.1.

In 1989 the United Nations General Assembly voted to hold the first United Nations Conference on Environment and Development (UNCED) in 1992 in Rio de Janeiro (the Earth Summit), involving '172 national representatives (of which 108 were heads of state) and over 2400 representatives of non-governmental organizations (NGOs)' (Taylor *et al.* 2005: 1680). While religious organizations were among the formally recognized NGOs attending the main summit, the event also hosted an 'alternative summit' (as did later conferences),

Figure 7.1 *The International Union for Conservation of Nature model of sustainability*

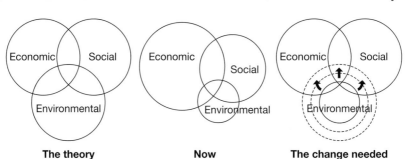

Source: http://cmsdata.iucn.org/downloads/iucn_future_of_sustanability.pdf

The three pillars of sustainable development: from left to right, the theory, the reality and the change needed to better balance the model.

which aimed to give an opportunity to individuals and groups, other than those from international governments and formal NGOs, to represent the 'grassroots' view. This reflected the widespread feeling that the interests of the vast majority of the world's population were not being represented by the 'official' bodies at these events. For example, at the parallel Rio summit there was participation from around 17,000 individuals, including representatives from Native American Indian communities, the Dalai Lama and a range of religious groups, involving 'Lutherans . . . Brazilian spiritualists. Hindu groups . . . and Tibetan Buddhists' (Fernandes 1992), who 'pressured the UNCED's official governmental representatives to strengthen their environmental commitments, with particular attention to marginalized peoples' (Taylor *et al*. 2005: 1681). Taylor *et al*. write that 'a strong religious voice promoting environmental sustainability and social justice emerged at Rio and continued to develop through subsequent United Nations environmental events' (ibid.: 1681). This has included the World Summit on Sustainable Development in Johannesburg in 2002 as well as the most recent, Rio +20, held in May 2012.

In the run-up to the 1992 Rio Earth Summit, discussion about a so-called Earth Charter was also initiated, calling for a set of norms to 'guide the transition to sustainable development'.[1] As Taylor *et al*. write, 'this initiative mimics the strategy that guided the United Nations Declaration on Human Rights, in which ratification by member-states is sought as a means to leverage better environmental behavior among individuals, institutions, and nation-states' (2005: 1682). It was launched in 2000 after a process of international consultation during the 1990s and has since 'been endorsed by over 2,500 organizations and institutions, including UNESCO and the World Conservation Union – IUCN' (Slaby and AtKisson 2007: 2). Moreover, representatives of religious traditions have been involved in the drafting of and subsequent promotion of the charter, and between 2007 and 2009 a project on 'religion and sustainability' was started within the Earth Charter Initiative that focuses 'on outreach to religious groups and leaders to seek their institutions engagement in using the Earth Charter . . . [as well as] research and the development of education materials designed specifically for religious audiences'[2] (see also Slaby and AtKisson 2007).

Thus, accompanying the international response to environmental issues at the governmental level, at least since the first United Nations international environment conference in 1972, there has been the emergence of a global environmental movement. This is a diverse social movement consisting of Northern and Southern organizations accompanied by a wealth of research and writing from different disciplines within the academy. Part of this response has been writing and social action from within religious traditions as well as from scholars and other commentators interested in the role that religions could play in environmentalism and sustainable development. Some scholars have used the term 'religious environmentalism' to describe this conscious and reflexive process of applying religious ideas to the issue of the destruction of the environment. They do so to distinguish a fundamentally modern phenomenon that arose alongside the emergence of the environmental movement in the

1960s, from the very ancient religious practice of worshipping features of the natural world – sometimes called 'nature religion' – which exists across many cultures.

Some examples of religious environmentalism, known as the Assisi Declarations, are presented in Box 7.2. In 1986 in Assisi, Italy, representatives of five of the world's major religions (Christianity, Buddhism, Islam, Hinduism and Judaism) met to make statements concerning the environmental nature of their religious traditions. This meeting was jointly organized by the WWF (the World Wide Fund for Nature) and the International Consultancy on Religion, Education and Culture (ICOREC), an organization based in Manchester. As the patron of the WWF, Prince Philip, the Duke of Edinburgh, attended this event. The Assisi Declarations comprised the first significant attempt by religious traditions to come together to discuss the relationship between their teachings and practices and the environment.

As we can see from the excerpts presented in Box 7.2, the core of religious environmentalism is the idea that religions teach that nature has a value beyond its use to humans and should be treated respectfully. It is the loss of this respectful attitude, it is argued, which has meant that humanity now treats nature as though it were an inexhaustible resource for its own use. Typically this is expressed in terms of the idea of nature being sacred, either because it has been created by God (e.g. in Christianity, Judaism and Islam) and therefore is a gift to humans that should be treated with respect, or because it is identical to the divine or imbued with the divine (e.g. in some versions of Hinduism). Buddhism, however, which does not hold with the belief in a supreme power, instead derives its ethic of care for nature from its teaching about compassion and non-violence (see Box 7.2). Moreover, there are a spectrum of approaches to religious environmentalism, from 'stewardship ethics' (which are popular in Christianity and promote the view that the Earth is a gift to humans from God, but a gift that they have a responsibility to treat respectfully) to more eco-centric or 'deep ecological' approaches. As is outlined in Box 7.3, 'deep ecology' considers that there is no moral difference between humans and nature and that humans should not act in such a way as to disrupt the natural balance. Some versions of Hinduism, for instance, which teach that both nature and humans are identical to God, are particularly well suited to expressing such eco-centric or deep ecological environmental ethics (see Box 7.2). By contrast, Abrahamic traditions (principally Christianity, Islam and Judaism) teach that God is separate from the Earth and humans, and therefore it is not possible to establish an ethic that places the Earth at the same level of importance to humans and the divine.

Thus, we can see that religious traditions have become drawn into environmentalism since the 1960s and appear to be important players in the modern environmental movement. Moreover, since the 'turn to religion', international development and environment organizations have become more likely to view religious environmentalism as relevant to their work on sustainable development. We will examine some examples of this in the final sections of the chapter.

Box 7.2

The Assisi Declarations

Excerpt from the Buddhist Declaration:

Buddhism is a religion of love, understanding and compassion and committed towards the ideal of non-violence. As such, it also attaches great importance to wildlife and the protection of the environment on which every being in this world depends for survival.

. . .From existing sources there is evidence to suggest that for all their limitations, people in the past were aware of this need for harmony between human beings and nature. They loved the environment. They revered it as the source of life and wellbeing in the world.

Excerpt from the Christian Declaration:

God created everything that exists, freely, by his word, and out of nothing. He alone is totally other, transcendent and immutable, whereas all creatures are contingent, mutable and wholly dependent on him for their existence. No creature can claim to be part of his nature or a 'spark' of his Being; but, by reason of its created origin, each according to its species and all together in the harmonious unity of the universe manifest God's infinite truth and beauty, love and goodness, wisdom and majesty, glory and power. . . .

[M]an's [*sic*] dominion cannot be understood as licence to abuse, spoil, squander or destroy what God has made to manifest his glory. That dominion cannot be anything other than a stewardship in symbiosis with all creatures.

Excerpt from the Hindu Declaration:

In the ancient spiritual traditions man [*sic*] was looked upon as a part of nature, linked by indissoluble spiritual and psychological bonds with the elements around him. This is very much marked in the Hindu tradition, probably the oldest living religious tradition in the world. . . .

[T]he natural environment also received the close attention of the ancient Hindu scriptures. Forests and groves were considered sacred, and flowering trees received special reverence. . . . The Hindu tradition of reverence for nature and all forms of life, vegetable or animal, represents a powerful tradition which needs to be re-nurtured and re-applied in our contemporary context.

Source: www.arcworld.org/downloads/THE%20ASSISI%20DECLARATIONS.pdf (from WWF 1986).

However, first it is necessary to ask where these religious environmentalist discourses are generated and by whom. In particular, there have been criticisms of religious environmentalist discourses for reflecting romantic western notions of a lost ecological idyll (an ecological 'golden age'), and critics argue that

Box 7.3

Deep ecology

The Norwegian philosopher Arne Næss coined the term 'deep ecology' in his 1973 article 'The Shallow and the Deep, Long-Range Ecology Movement: A Summary'. Næss considered that mainstream environmental movements had tended to concentrate upon the symptoms of the environmental crisis rather than its causes. In lobbying for legal and political solutions to the destruction of the environment, they had achieved little more than applying a 'band aid' to the problem and environmental devastation was only being postponed. This 'shallow ecology' was contrasted with a deep ecological approach that called for a transformation in the way in which humans relate to nature. Deep ecology is opposed to a mechanistic worldview that values the natural world because of the resources it provides to serve human ends. Instead it insists that all life is interconnected and has equal as well as intrinsic value. This 'eco-centric' ethic supports a holistic worldview that challenges anthropocentric attitudes towards nature.

Although deep ecology did not emerge until the early 1970s, it draws upon earlier movements and traditions, including romanticism, eastern religions such as Taoism and Buddhism, as well as the writings of individuals such as Henry David Thoreau, Aldo Leopold and Baruch Spinoza. However, the extent to which deep ecology is a consistent body of thought is debatable. Its sources are diverse and it is a product of emotive as well as rational thought, with adherents commonly claiming that their belief in deep ecological values is a result of a personal and even quasi-sacred relationship to the Earth. Deep ecology evolved from the experiences of those engaged in environmental direct action, the most prominent manifestation being the Earth First! movement that emerged in the early 1980s in the United States.

While the action orientation of deep ecology can be seen as a strength, its system of environmental thought has been challenged for being irrational and mystical, as well as implicitly anti-people. Critics perceive that its Malthusian stance on population and its emphasis upon wilderness preservation work to the disadvantage of the poor in developing countries. A high population growth rate is blamed for environmental destruction and the poor are to be evicted from their lands in an attempt to preserve designated areas of wilderness. Moreover, critics ask, 'Why should we agree that all life has equal value?' Where deep ecology considers that this idea of biocentric equality reflects a non-anthropocentric attitude towards the natural world, it fails to realize that this ideology is itself the product of human constructions.

Source: adapted from Tomalin (2004b).

these discourses are more likely to be adopted by an educated 'middle class' that has been influenced by western eco-centric and romantic approaches to environmentalism. In the following section we will explore this critique and will assess the extent to which religious environmentalism has relevance for the poor who are struggling against floods, famines and droughts. This discussion will mainly focus on debates about religious environmentalism in India and with respect to the Hindu tradition, although the arguments presented are relevant to other contexts, and some examples will also be given from other locations and religious traditions.

Critiques of religious environmentalism and the 'myth of primitive ecological wisdom'

In tandem with literature that provides environmental interpretations of religious and cultural traditions, there have emerged critiques of such exegeses. Some critics argue that religio-cultural traditions can also be interpreted to contradict contemporary environmentalist thinking (Harris 1991; Nelson 1998). For instance, Nelson is concerned that the frequent disregard for the material world in Hinduism, as an impediment to spiritual progress, is problematic for claims that the tradition is environmentally friendly. He questions the sentiments towards nature that a passage such as the following might suggest: 'Pure non-attachment is disregard for all objects – from the god Brahmā down to plants and minerals – like the indifference one has toward the excrement of a crow' (from the *Aparokshanubhuti*, a text associated with the eighth-century Indian philosopher Shankara; cited in Nelson 1998: 81).

Nelson continues by asking: would this inspire the devotee to revere nature as part of a spiritual life, or 'would it rather teach . . . the irrelevance of nature to spiritual life?' (1998: 81). Other critics have argued that environmentalist interpretations of religion are anachronistic, that they read contemporary concerns into religious or cultural traditions (Freeman 1994; Pedersen 1995). While there is often the view in religious environmentalism that ancient peoples worshipped elements of the natural world because they were aware of their ecological value, we cannot assume that that was the case. Many religious traditions can be interpreted to support contemporary environmentalist thinking, but this does not mean that people who belong to those traditions – either in the past or today – will necessarily think and behave like modern environmentalists. For instance, whereas modern environmentalists consider that the whole of nature is relevant and worthy of respect, many examples of nature religion are focused only on certain aspects of the environment, for instance particular trees, forests or rivers.

In India, for example, within the Hindu tradition we find many examples of the worship of elements of the natural world that do not necessarily result in behaviour that is directed towards environmental conservation. For instance, while people in India worship the dangerously polluted River Ganges as the goddess Ganga Ma, there is little evidence that this religious practice motivates devotees to engage in initiatives to prevent any further pollution of the river (Alley 1998, 2000, 2002; Plate 7.1). Similarly, there is a strong tradition of sacred grove preservation in India, yet it can be suggested that people worship these forests because they are the abode of the deity rather than to conserve biodiversity (Freeman 1994; Tomalin 2004a). The eco-centric or deep ecological

Plate 7.1 *A Hindu ritual offering to the Ganges, Varanasi*

Source: J. Duval (Creative Commons). Online, available at:
http://commons.wikimedia.org/wiki/File:Offering_to_the_ Ganges,_Varanasi.jpg?uselang=en-gb.

view that nature has intrinsic value and humans ought to put the 'Earth first' cannot be inferred from these examples of nature religion. Instead, it has been widely argued that the idea of the 'intrinsic value' of nature (which non-industrial or non-western people are presumed to uphold) has its roots in the European Romantic movement (itself a reaction against industrialization), which idealized nature as a realm of purity and aesthetic value (Tomalin 2009c; Guha 1989).

In sub-Saharan Africa also there are increasing numbers of scholars and activists who promote the role of religious environmentalism, arguing that western models of development as well as solutions to the problems created by industrialization should be rejected in favour of drawing upon traditional worldviews and epistemologies, shaped by what is known as African traditional religion (ATR) (Dikirr 2008). Taringa, for example, writing about the Shona in Zimbabwe, states that 'African traditional religion, and Shona religion in particular, is generally regarded to be intrinsically environmental friendly' (2006: 191). However, these beliefs are often portrayed as being threatened by the impact of Christianity and westernization. He argues that such perceptions are idealistic and romantic, and need to be reconsidered, 'especially so if Shona religion is to re-emerge as a stronger environmental force in the global village' (ibid.: 192). He suggests that Shona attitudes towards nature are both 'ecologically responsible and harmful' (ibid.: 213). For instance, 'not all animals, plant life and water sources are sacred. This means not all aspects of nature play a pivotal and vital role in their beliefs about salvation' (ibid.: 211). Moreover, since the colonial period the rapid changes towards a cash economy and modern ways of farming, as well as the emergence of district councils parallel to the traditional system of village chiefs, make it hard to view such romanticism as practicable (ibid.: 212). In another study, Dikirr similarly concludes that environmental movements that focus on rediscovering 'cultural roots' do not offer a viable option for Africa's renaissance, even though they emerge as a response against the perceived problems associated with western models of development, also viewed as an extension of the colonial project. Instead, he argues that 'what scholars and students of Africa need to do . . . is to figure out how a creative synthesis between Africa's village-level of cooperative ethic, system of local governance, indigenous technologies of managing natural resources and Western techno-scientific skills might be realized' (ibid.: 96).

Thus, a number of scholars have been critical of the ways in which many modern environmentalists assume that indigenous peoples, tribal communities or non-industrial societies are inherently inclined to hold eco-centric views (Baviskar 1997). This has been called the 'myth of primitive ecological wisdom', which articulates the desire for a return to a mythic state of ecological harmony, often drawing upon potent images of the sacredness of the Earth (Milton 1996; Tomalin 2009c). Guha, for instance, one of the key commentators on this topic in India, is concerned that there is a 'persistent invocation of Eastern philosophies as an antecedent in point of time but convergent in their structure with deep ecology' (1989: 76). He criticizes this as an inherently orientalist discourse, where 'Eastern man [*sic*] exhibits a spiritual dependence with respect to nature . . . [that has] the characteristic effect . . . of denying agency and reason to the East and making it the privileged orbit of Western thinkers' (ibid.: 77). Guha argues that this deep ecological romanticization of eastern traditions is part of a broader process of western orientalist construction: this appropriation of eastern cultural resources by the West tells 'us far more about the Western commentator and his desires than about the "East"' (ibid.: 77).

Others have drawn attention to the ways in which this 'new traditionalist discourse' (Sinha *et al.* 1997) has become part of a particular type of environmentalism in India, of the 'bourgeois' or middle-class variety (Mawdsley 2004: 87). Within this 'new traditionalist discourse', Baviskar locates what she calls the 'Hindu civilizational response' of middle-class scholars and activists who appropriate the Hindu tradition as a source of ecological ethics and wisdom (1999). She critiques this as a form of 'inverted orientalism' for 'positively valorizing a mythical Other created by Western environmentalists' (ibid.; Guha 1989; Inden 1986). However, Baviskar juxtaposes the 'bourgeois environmentalism' of the middle and upper classes in India and the survival-related ecological concerns of the lower-class peasants and tribals (1997). The 'bourgeois environmentalist' has typically been exposed to western, globalized environmentalist discourses (including deep ecology/eco-centrism), is affluent and is more likely to be concerned with nature for its leisure potential and aesthetic value (Baviskar 2003). Thus, it is not the case that poor people are unconcerned about environmental issues; rather, they are likely to be concerned for different reasons and in different ways. Guha and Martínez-Alier (1997) make a similar distinction between the 'empty belly environmentalism of the South' and the 'full belly environmentalism of the North'. They distinguish between 'quality of life' and 'survival needs' as motivating factors indicating different sources of environmental concern (ibid.). An example of the potential for tensions between 'bourgeois' and 'survival-related' environmentalisms is the creation of national parks in Africa and Asia. This is explored in Box 7.4 with respect to Africa.

Thus, while religious environmentalists have become increasingly vocal within contemporary debates about biodiversity conservation or climate change, others are critical, saying that religious environmentalism amounts to the selective appropriation of religious beliefs and practices that support the idea of nature as sacred and that it tends to romanticize many groups of people in developing contexts as inherent environmentalists. However, a number of commentators, while agreeing with the importance of this critique, are also keen to point out that the situation is actually more complex. As Milton notes, the 'myth of primitive ecological wisdom is not simply a notion imposed by romantic environmentalists on a sector of the world's population, [but] it is also an image which indigenous peoples accept and promote for themselves' (1996: 202). Brosius is also critical of the tendency to portray as a 'myth', the idea that indigenous peoples live, or lived, in harmony with nature. He argues that it can

---

Box 7.4

*National parks: tensions between 'bourgeois' and
'survival-related' environmentalisms*

Nelson argues the nineteenth-century image brought back to Europe by explorers and missionaries of the 'noble savage' and of an Africa that was 'a virtual Garden of Eden, innocent of the ills of modern civilization' (2003: 70) has been recreated in the form of the national park. However, the 'Eden myth [which] Africans present has been supplanted by images of an Edenic wilderness in which current Africans, as well as non-Africans (except tourists), must be excluded' (ibid.: 70). He writes:

> The greatest current efforts to 'save' Africa are associated with contemporary environmentalism. The results have not been as devastating as the experience of slavery, yet they have often served Western interests and goals much more than the interests of ordinary Africans. In some cases, local populations have been displaced and impoverished in order to create national parks and to serve other conservation objectives. Under the banner of saving the African environment, Africans in the last half century have been subjected to a new form of 'environmental colonialism'.
>
> (ibid.: 65)

---

undermine the integrity of indigenous communities who adopt this 'myth' as part of the process of securing their autonomy and defining their interests, since it plays into the hands of their opponents, who can then easily dismiss indigenous people's environmentalist discourse as fabrication (1999). He argues that we need to distinguish between 'strategic' and 'romantic' essentialisms, since 'historically marginalized communities have begun to recognize the political potency of strategically deployed essentialisms' (ibid.: 281; see also Spivak 1987; Fuss 1989). Therefore, in assessing the relevance of apparent romantic styles of religious environmentalism in different contexts, it is also important to consider the extent to which strategic essentialisms are at play. People sometimes adopt essentialist discourses in order to construct and maintain shared identities within oppositional socio-political movements. However, there is also a danger that strategically employed essentialisms can be so 'successful', seductive and persuasive (particularly if they reinforce stereotypes held elsewhere) that they become potentially risky in that they may come to shape how other constituencies represent those captured by the discourses.

Concern has been expressed that the 'myth of primitive ecological wisdom', which often includes romantic visions of the ways in which people's religion contributed towards ecological wisdom, has had an influence on some of the ways in which mainstream development actors approach the theory and practice of sustainable development and that this may result in initiatives or approaches that actually undermine development goals. One area where this has been explored in the literature is with respect to the ways in which understandings of women's relationship to the environment have been influenced by the 'myth of primitive ecological wisdom' in the form of 'ecofeminism'. Moreover, while ecofeminism has been ardently rejected by many feminists for what is perceived to be crude biological determinism and essentialism about women's relationship towards nature, others have also viewed it as an example of an ecological approach that employs such essentialisms strategically (MacGregor 2007; Sturgeon 1997). However, this view is risky, since these strategic deployments may actually come to shape the ways in which women are viewed within other constituencies, including development agencies, which play a role in shaping the material conditions of women's lives. In the following subsection we will examine ecofeminist discourse as a manifestation of religious environmentalism and the ways in which it influenced the 'women environment development' (WED) approach within mainstream development agencies during the 1980s and 1990s.

Ecofeminism and the WED approach

While there are many strands within ecofeminist thinking (Warren 1996), the underlying position is summed up by Green *et al.* (1998). They write that ecofeminists critique the dominant model of development, which is perceived as a male construct, and one that has promoted economic development in ways which have been

> harmful both to women and to the environment by trampling alternative, local knowledge, especially women's knowledge, associated with organic concepts of people and nature as interconnected; by disregarding the spiritual and sacred in people's attitude to their environment and women's special role therein; and by overriding holistic and harmonious environmental practices. . . . [E]cofeminists consider that the feminine principle is not quite extinct in the environmental context, but still manifest in a residual, near instinctual wisdom which some women have been able to retain in the face of developmental pressures. . . . '[T]hird World women' are portrayed as the last bastion of feminine environmental wisdom and they provide the key to its retrieval.
>
> (ibid.: 273)

Thus, women are considered to have a spiritual connection with the natural world, one that has contributed to their inclination to protect nature, and this is severed when ecosystems are destroyed. Within the range of types of ecofeminism, 'religious' or 'cultural' ecofeminists emphasize these links most strongly and draw attention to various expressions of nature religion involving the worship of forms of the feminine divine considered to be immanent in nature (e.g. manifestations of the Mother Earth Goddess) as sources of women's empowerment as well as strategies for environmental protection. They are concerned that patriarchal religion, modes of production and technological developments have been devastating not only for nature but also for women. For instance, traditional Christianity teaches that nature, humans and the divine are distinct entities ranked hierarchically. However, the separation of God from nature has meant not only that humans are more likely not to treat nature respectfully, but also that women's identification with nature (e.g. through their capacity to reproduce) causes them to be seen as entities to be dominated and controlled by men.

With respect to the Hindu tradition, the Indian ecofeminist Vandana Shiva makes this link explicit in her writing when she argues that

> women in India are an intimate part of nature, both in imagination and in practice. At one level nature is symbolised as the embodiment of the feminine principle, and at another, she is nurtured by the feminine to produce life and provide sustenance.

> (1988: 38)

Shiva considers that women are closer to nature because nature (*prakriti*) is seen as feminine in the Hindu tradition. However, she also points out that nature (*prakriti*) in Hinduism is considered to be a goddess of equal standing to the male god. For Shiva, the introduction of western modes of development into the 'Third World', particularly intensive agriculture, is a form of 'mal-development' resulting in the 'death of the feminine principle' and as such is at odds with the traditional Hindu worldview that elevated the goddess. DasGupta Sherma (1998), however, has criticized Shiva for adopting a particular reading of the Hindu tradition, since not all versions consider nature and the goddess in this way, with others viewing nature/*prakriti* as something to be escaped from, since it binds the self, or soul, to future rebirths. However, many ecofeminists similarly re-evaluate what they see as a myth of patriarchal progress and instead envisage a return to small-scale agricultural communities that worship the Earth goddess, communities in which women's natural inclination to work with nature rather than against it is permitted to flourish. Generally they view non-western religious traditions as

> free of the nature–culture dualism which is believed to underpin the oppression of both women and nature in Western history and thought, and the absence of which is thought to engender positive and sustainable relations between peoples and their environments in many non-Western societies.

> (Jackson 2001: 23; see also Merchant 1992;
> Plumwood 1986; Warren 1987)

However, these discourses about women's 'primitive ecological wisdom' within ecofeminism have not been confined to the fringes of the feminist movement but have instead, as Leach suggests, shaped the 'women environment development' (WED) approach within donor-driven development policy and practice since the 1980s, a decade that witnessed a rise in global environmental concern. She writes:

> The woman head-loading firewood across a barren landscape has become an environment and development icon. Reproduced in policy reports, NGO glossies and academic books alike, her image encapsulates powerful and appealing messages. . . .

> These material dimensions were bolstered by fables about women's natural, cultural or ideological closeness to nature; varieties of 'earth mother' myths which could be, and were, used to justify women's roles, as well as to give cultural and political appeal to the notion of global environmental sisterhoods.
>
> (2007: 67–68)

Prominent ecofeminists such as Vandana Shiva were present at the 1992 Rio Earth Summit and, as Leach writes, influenced 'the Miami declaration adopted by a large international conference of women activists prior to the United Nations Conference on Environment and Development (UNCED) conference in Rio, 1992' as well as the 'preamble to Women's Action Agenda 21 discussed at Rio, which linked the highly specific experiences of diverse groups of women in localized environmental protection with a broad critique of mainstream economic and military processes' (2007: 71). While the view that ecofeminism was a strategic tool that enabled women to gain access to high policy meetings like the Rio Summit is appealing, we also need to consider the longer-term effects (MacGregor 2007: 51) In particular, as Leach writes,

> [i]n this context, it is not surprising that echoes of ecofeminist discourse crept into the statements of donors and NGOs associated with much less radical visions. Together, ecofeminism and WED supported a view that agencies should identify women as allies – or even the prime movers – in resource conservation projects.
>
> (2007: 71)

WED interventions typically involved an emphasis on women-centred environmental/conservation projects, including women-focused social forestry, agroforestry, soil and water conservation projects, fuel-efficient stoves and solar cookers (Buchy and Rai 2007). However, as Jackson (1994) points out, these projects tended to treat women's time and labour as flexible and inexhaustible, and often just added to their other tasks. Fuel-efficient stoves, for instance, created more work for women, as they need tending and are unsafe for small children (ibid.). Moreover, there was an assumption that women are naturally in favour of 'sustainable development', when in fact women in many cases had benefited from the Green Revolution. WED has been criticized for treating traditional gender roles as natural and for rarely involving women in decision-making processes. Within this discourse, women were commonly depicted as bearing the brunt of environmental problems but also as having a 'natural' inclination towards environmental conservation.

However, from the early 1990s these ecofeminist-influenced WED perspectives began to be critiqued, and it also seemed to be the case that some of the donors who had previously promoted these myths were themselves beginning to rethink them (Leach 2007). While not wanting to diminish the importance of attempts by environmental agencies to address gender, Leach (herself one of the critics) writes, '[i]t seemed appalling and dangerous that this was occurring through an approach with glaring flaws influenced by dubious ecofeminist work' (ibid.: 73). Writing in the mid-2000s, she concludes that such myths do seem to have receded and it is no longer *de rigueur* to depict women as 'natural carers' (although she does warn that 'there is a danger that the baby has been thrown out with the bathwater; gender-blind environment and development work seems to be on the rise' (2007: 68)). The alliance between WED and ecofeminism has had its moment, spurred by the moral panic underlying the Rio Earth Summit in 1992 and the pressure on development agencies to do something about environmental sustainability, and now 'ecofeminist fables seem largely to have retreated back into the world of academic writings and fringe environmental groups which originally spawned them' (ibid.: 82). Nonetheless, even in the 2000s the hegemony of these discourses about women and nature has been difficult to shake off completely, with the United Nations Environment Programme's manual *Women and Environment* (UNEP 2004), for instance, blatantly promoting the tapping of women's and local people's productivist and participatory potentials to ensure efficient and supposedly sustainable use of resources (see Plate 7.2 for the style of image being critically referred to).

WED approaches did not incorporate ecofeminist discourses because of their link to religion; instead, ecofeminism was seductive because its discourses presented a way of thinking about women's relationship to nature that appeared to be rooted in tradition and culture (and therefore authentic). Nonetheless, the lessons learned from this example are relevant for more recent initiatives within mainstream development policy and practice that engage with religion in pursuit of sustainable development. The shaping of WED approaches by ecofeminist discourses during the 1980s and 1990s suggests that development actors need to be careful not to collude with the 'myth of primitive ecological wisdom', which may be articulated as a strategic essentialism in some contexts but could be damaging if taken to represent the ways in which people actually relate to their natural environments.

Bearing this warning in mind, and having examined the provenance of religious environmentalism, we will now return to the questions posed at the outset concerning the role that religions could play in achieving MDG 7, which is concerned with sustainable development: *What role might religion play in biodiversity conservation, the lobbying of governments and industry to adopt more sustainable policies and practices, and appealing to individuals to change their behaviour where it is contributing to environmental problems, in order to contribute positively towards improving people's livelihoods and well-being? Could religion in developing countries have a positive impact on reversing climate change and preserving species, in the interests of people's long-term development, or is this a 'post-materialist' romantic myth that essentializes poor people as inherently environmentalist?*

Plate 7.2 *Woman in Mysore balancing a basket of* chikku *(or* sapota – a type of fruit) *on her head*

Source: Wen-Yan King (Creative Commons).

Religious traditions and sustainable development: some examples of international initiatives

The critique that religious environmentalist discourse romanticizes the links between interpretations of tradition and what people actually do (in the past and today) and that it essentializes poor, 'religious' people as inherently close to nature (therefore making a range of assumptions about what is ultimately in their best interests) ought to be taken seriously. Nonetheless, there is a wide range of examples of both grassroots and international environmentalist initiatives globally that draw upon religious belief, as well as the local networks and legitimacy of 'faith-based organizations'. There is some evidence that these are increasingly taken notice of by mainstream development actors, such as the World Bank and the United Nations Development Programme (UNDP), and in this section we will explore several international initiatives that engage with religion with the aim of securing more environmentally sustainable practices and outcomes, as key to sustainable development.

We will begin the discussion with an overview of the work of an international NGO called the Alliance of Religions and Conservation (ARC), which supports projects across the globe that combine religion with conservation, and which has also gained the support of the World Bank and the UNDP for its activities.

The Alliance of Religions and Conservation (ARC)

ARC describes itself as:

> a secular body that helps the major religions of the world to develop their own environmental programmes, based on their own core teachings, beliefs and practices.
>
> We help the religions link with key environmental organisations – creating powerful alliances between faith communities and conservation groups.[3]

ARC grew from the ideas and interest generated by the meeting held in September 1986 in Assisi when representatives of five of the world's major religions (Christianity, Buddhism, Islam, Hinduism and Judaism) presented declarations concerning the environmental nature of their religious traditions. Following this meeting, the Network on Conservation and Religion, of which WWF was the primary sponsor, was formed and a magazine called *New Road* (this was later replaced by another publication, *News from ARC*) was launched to publicize its activities and to present articles on different aspects of religions and environmental conservation across the globe. This publication contained articles about people inspired by their religious traditions to undertake environmental projects, as well as about the ecological content of religious traditions and stories about environmental disasters or the destruction of 'ecologically benign cultures'. By 1995, Baha'is, Jains, Sikhs and Taoists had joined the alliance, and representatives of a total of nine faiths were invited to meetings in Japan and at Windsor Castle in the United Kingdom for the 'Summit on Religions and Conservation'. At these meetings the Alliance of Religions and Conservation was launched and a nine-year plan agreed for each tradition. 'For the first time an organisation existed to link the secular worlds of conservation and ecology with the faith worlds of the major religions.'[4]

According to the website of the organization, ARC specifically aims to encourage people to 'tread more gently on the earth' through realizing that care for nature is at the core of their religious traditions, and the website includes an overview of religious attitudes towards nature from different traditions.[5] In addition to ARC's harnessing of the role that religious ideas and actors can potentially play in biodiversity conservation, projects it supports are also involved in the lobbying of governments and industry to adopt more sustainable policies and practices, encouraging people to change their behaviour where necessary (e.g. to undertake recycling) and promoting the use of renewable energy (Plates 7.3 and 7.4).[6]

In addition to working at the grassroots level, ARC has also been successful in courting the support of major influential development bodies, such as the World Bank, and United Nations agencies, such as the UNDP.

Plate 7.3 *Group photograph of participants at an ARC-organized conference held in Nairobi in March 2011*

Source: ARC: the Alliance of Religions and Conservation (Creative Commons). Online, available at: www.flickr.com/photos/53990852@N05/7886204706/in/photostream/.

Plate 7.4 *Archbishop Abuna Gregorios of the Ethiopian Orthodox Tewahedo Church attending a workshop on biogas and sustainable farming methods, 18 September 2012*

Source: ARC: the Alliance of Religions and Conservation (Creative Commons). Online, available at: www.flickr.com/photos/53990852@N05/7886112404/in/photostream/.

In 1997, ARC arranged a meeting at Lambeth Palace, the residence of the Archbishop of Canterbury, between leaders of the nine faiths participating in its work and the then president of the World Bank, James Wolfensohn (also the founder of the World Faiths Development Dialogue in 2000): 'The aim was to discuss ways in which alternative economic models arising from the faiths could help reduce poverty and environmental destruction.'[7] According to the ARC website, emerging from this meeting were 'stronger links' between ARC and the World Bank, including joint projects and a book entitled *Faith in Conservation* (Palmer with Finlay 2003).[8] In 2006 the World Bank published a report, *Faiths and the Environment*, outlining its work with ARC, and other organizations, in this area. Indeed, as the World Bank states on its own website:

> While the world's major religions have been, until recently, relatively voiceless in the environmental debate, it is being shown that they can represent a very powerful voice for environmental stewardship.
>
> This is why the World Bank cooperates with the major faiths as partners. . . .

The 11 faiths that now make up ARC represent two-thirds of the world's population. They own around 7 percent of the habitable surface of the planet, they have a role in 54 percent of all schools, and their institutional share of the investment market is in the range of 6–8 percent. They are serious stakeholders in development.[9]

In addition to the work with the World Bank on poverty reduction and biodiversity conservation, in December 2007 the UNDP launched a three-and-a-half-year project to work with the world's major religions to tackle climate change. This has been jointly managed with ARC and has involved faith traditions in drawing up seven year plans, to run from 2009 to 2016. The plans were launched at an event in 2009 involving FBOs working on development and environment issues such as CAFOD and the Bhumi Foundation (an international Hindu environmental organization) as well as representatives of particular religious traditions from around the world including Mongolian Buddhists, Armenian Christians and Cameroonian Presbyterians:

ARC will help representatives of the faiths draw up their programmes and implement them throughout the world – from Sikh gurdwaras in India which provide food for millions of poor people every day, and which are aiming to introduce ecologically sustainable fuel sources to their kitchens, to Buddhist ecology centres in China, to Ethiopian monasteries teaching organic farming practices, to Catholic programmes to preserve Latin American forests, to Muslim fishermen protecting the seas off the coast of East Africa.[10]

The above overview is mainly derived from the website of ARC, which provides little detailed critical evaluation of the outcomes of projects that combine religions and ecology over both the short and the long term, or about the outcomes of its engagement with the UNDP and the World Bank. Whether, for instance, the will or the resources will be available to evaluate the outcomes of the individual plans submitted as part of the UNDP–ARC project on climate change remains to be seen. Moreover, the fact that only two publications were produced with the World Bank does not suggest that this topic has been given the attention and focus that it probably deserves. Nonetheless, the promotional nature of this material is understandable since the role of a website of an organization is not to provide critiques of, or to assess the limitations of, its work. However, studies that undertook this type of assessment and evaluation of outcomes would enable us to arrive at a better understanding of the relationships between religions and environmentalism, and their significance for sustainable development. A greater emphasis upon critical research in this area is undoubtedly necessary, such that religion can be incorporated into environmental projects in ways that do not replicate and reinforce dangerous and unhelpful romanticized stereotypes, but instead contribute positively to improving people's livelihoods and well-being.

In the remainder of this chapter I will critically evaluate one area where the links between religions and environmental sustainability have received a significant level of attention, namely with respect to the links between the preservation of sacred groves in many parts of the globe and biodiversity conservation. However, to what extent does the focus on these areas reflect the yearnings of romantic religious environmentalists, based on assumptions about the role that religion has and can play in environmental conservation? Or do discourses about sacred groves actually have the potential to achieve biodiversity conservation in some contexts, as well as improving livelihoods and well-being?

Sacred groves, biodiversity conservation and sustainable development

In contrast to other environmental issues, such as the need to consume less, adopt more environmentally friendly farming methods and reduce the ecological impact of transport or construction, biodiversity conservation may seem comparatively less important when it comes to thinking about improving livelihoods and well-being. This is particularly the case since there has been some concern that the protection of biodiversity is often at odds with development goals when it seeks to limit or to prevent people's interaction with the environment, as in the case of national parks (see Box 7.4 on p. 187). However, across the globe both small local NGOs and large international bodies recognize the close links between biodiversity conservation and poverty reduction. According to USAID:

> Biodiversity conservation, the practice of protecting and preserving the wealth and variety of species, habitats, ecosystems, and genetic diversity on the planet, is important for our health, wealth, food, fuel, and services we depend on. It plays an integral role in supporting many sectors of development.[11]

In terms of the role of religion in protecting biodiversity, the preservation of sacred spaces, including sacred groves, in many parts of the globe has received attention and has become the focus of a significant body of literature as well as the basis of a range of initiatives. Box 7.5, for instance, provides a list of international organizations which recognize the role that the protection of sacred spaces might play in biodiversity conservation. This is not, however, just a debate about biodiversity conservation for its own sake; it also draws attention to the protection of sacred groves as part of local community methods of forest management, methods that are relevant to thinking about sustainable development in different locations. Moreover, to view the preservation of sacred spaces, which are sometimes threatened by modern development projects and initiatives, as important in itself is relevant to conceptions of sustainable development that go beyond a focus on economics. UNESCO, for instance, has been involved in initiatives to protect sacred groves or forests in different parts of the globe, recognizing the importance of preserving cultural diversity as an end in itself but also seeing that doing so may contribute towards biodiversity conservation. However, before we examine UNESCO's work in this area we

---

Box 7.5

*International organizations that recognize that the protection of
sacred spaces may be relevant for biodiversity conservation*

- Conservation International – Faith-Based Initiatives Program;
- the Earth Island Institute – Sacred Land Film Project;
- the International Council on Monuments and Sites (ICOMOS);
- the Man and the Biosphere Programme (MAB) – World Network of Biosphere
  Reserves, a unit of UNESCO;
- the Mountain Institute – Sacred Mountains Project;
- Sacred Sites International;
- the World Conservation Union, formerly the International Union for the Conservation
  of Nature and Natural Resources (IUCN), with its many different groups and their
  projects such as the World Commission on Protected Areas (WCPA) and the Delos
  Initiative;
- the World Heritage Centre and World Heritage Sites of UNESCO;
- the World Wildlife Fund, now called World Wide Fund for Nature (WWF). The latter
  has been collaborating with the Alliance for Religions and Conservation (ARC).

Source: www.eoearth.org/article/Sacred_places_and_biodiversity_conservation (last accessed 11 September
2012).

---

will first look at critiques of the idea that sacred groves have contemporary
ecological significance on the basis that this is an assumption that is part of
romantic religious environmentalist discourses.

Critiques of modern ecological discourses about sacred groves

Sacred groves exist across the globe and are often seen as the abode of divine
powers, which can afford them protection from human interference. Some
include areas of pristine forest and constitute 'biodiversity hotspots', whereas
others may have undergone human interference or may consist of little more
than a few shrubs. However, Barre *et al.*, writing about sacred groves in Ghana,
tell us that 'spiritual beliefs and practices (taboos) play a pivotal role in the
conservation of sacred natural sites' (2008: 31). These groves have declined
significantly in size, quality and number under the influence of encroachment
in recent years due to increased human need for land and forest resources, as
well as changing religious practices. For instance, in India it has been suggested
the dominance of Brahmanical Hinduism has undermined localized indigenous
practices such as those relating to sacred groves. In particular, this process of
'Sanskritization' has meant that some encroachments may be carried out 'on
behalf of the divine' (Kalam 1996: 51), in order to build temples which the
Brahmanical gods prefer (Gadgil and Chandran 1992: 187; Chandran and
Hughes 1997: 420). In Africa, many consider that the coming of Christianity
has meant that people have less respect for sacred groves that were preserved
as part of African traditional religion (Barre *et al.* 2008).

There have, however, been critics of discourses about the environmental significance of sacred groves who draw attention, first, to the idea that they are not necessarily evidence of 'environmentalism', and second, that there is not an ideal type of sacred grove that is constituted by 'pristine forest' (Freeman 1994; Kalam 1996). Writing about sacred groves easily lends support to religious environmentalist discourse, where practices such as preserving sacred groves are seen as evidence of a 'primitive ecological wisdom' that pre-dates colonial interference in natural resource management. However, it seems more likely that any protection of biodiversity was coincidental rather than intentional and that sacred groves were protected out of respect for the deity rather than because of an innate belief in the intrinsic value of nature. As Kalam points out, 'where [in earlier times] was the need to preserve/conserve the forests as there was no danger at all of depletion of forest resources?' (1996: 52). Moreover, the protection of sacred groves does not match modern ecological thinking, as it only affords protection to particular areas of nature rather than the entire natural environment.

Freeman, whose work focuses on India, has also argued against the tendency of much discourse around the subject of sacred groves to assume, or at least to give the impression of, the existence of an ideal type of sacred grove as 'pristine forest' (1994: 9, n. 11). He suggests that sacred groves have always taken a variety of forms that do not necessarily coincide with the modern environmentalist's idea of 'pristine relics from a primeval past' (ibid.: 11). Moreover, a 'kavu [sacred grove] may refer to a temple that no longer has any associated grove, [thus] the semantic weight of the term rests with the dedication of the site to a deity, rather that with the flora of the site, per se' (ibid.: 12, n. 14). The purpose of sacred groves is to serve as a pleasure garden for the use of the deity and, while encroachment according to human needs is forbidden, it is often the case that 'what we might regard as human disturbance, resource exploitation, and encroachment are happily accommodated within the cultural framework of the grove as the deities' preserve' (ibid.: 11).

While these criticisms are valid, in the following subsection we will explore the extent to which some sacred groves might play a role in modern strategies to conserve biodiversity in ways that have relevance for sustainable development.

The protection of sacred natural sites and biodiversity conservation

UNESCO has been instrumental in protecting World Heritage Sites across the globe since the adoption of the World Heritage Convention in 1972. In 1992 the category of 'cultural landscapes' was added, paving the way for the inclusion of sacred natural sites on the list of Word Heritage Sites. Today these include the Sacred Mijikenda Kaya Forests in Kenya (Githitho 2003) and the Osun-Osogbo Sacred Grove in Nigeria. A report published in 2003, drawing on the proceedings of an international workshop organized by UNESCO, 'The

Importance of Sacred Natural Sites for Biodiversity Conservation', outlines case studies that focused on 'traditional mechanisms of environmental conservation in the different regional contexts of Africa, Asia-Pacific and Latin America' (Lee and Schaaf 2003: vii). In this report, Schaaf asks:

> Can cultural values and traditional belief systems, which respect the environment, be a more powerful or at least an equally powerful means to conserve nature than legally protected areas? Such an approach appears promising in many traditional societies of the world where the concept of a 'sacred nature' is well embedded in societal norms.
>
> (ibid.: 18)

A SWOT analysis of the evidence presented at the workshop suggests that 'strengths and opportunities' include the high conservation value of many sacred natural sites (SNS); the added value to conservation of shared community beliefs about preserving SNS when compared with legal regulation alone; the fact that they often preserve traditional knowledge; and the potential they offer for developing eco-tourism. 'Threats and weaknesses' include the fact that they are often not recognized in national policies and legal systems, and the local knowledge, with which they are intrinsically linked, is not valued as highly as modern science in official circles; and that SNS are not necessarily areas of pristine wilderness with high levels of biodiversity and they typically have a function beyond conservation, with the potential for a clash between conservation priorities and the religious use of these places (ibid.: 163–165).

Reflecting these 'threats and weaknesses' more broadly, there has been some criticism in the literature of this conservationist preservation of sacred groves for being instrumentalist, and, as Engel writes, 'most positive evaluations of the ecological role of traditional beliefs in preserving sacred places are advanced for scientific and social purposes that are extrinsic to the beliefs themselves' (2005: 193). Moreover, it cannot be assumed that the goals of scientific conservation will necessarily match the cultural and economic needs of local communities, particularly where these deny any human interference whatsoever. As we have seen, Freeman noted that different degrees of human interference in sacred groves may well be accommodated with the 'cultural framework of the grove as the deities' preserve' (1994: 11). In the course of their research in Ghana, Barre et al. heard 'reports of animosity between local people and protected areas because of the exclusionary approach to protected area management' (2008: 37). They continue that while 'sacred groves are endogenous protected areas which gives them a margin of community acceptance . . . [to] conserve the groves in tandem with local communities, conservators must resist the temptation to annex the groves for "proper" conservation' (2008: 37). Although the majority of respondents in their study area were accepting of the need for outside assistance in conserving the groves, they were also clear that 'their religious and cultural beliefs should not take second place' (ibid.: 37). Barre et al. warn that:

[o]ften, researchers romanticize culture and assume that once the objective is to preserve culture there will be zero opposition from local communities. Failure to grasp the centrality of religious and cultural beliefs for locals could derail well-meaning, potentially mutually beneficial efforts.

(ibid.: 37)

Nonetheless, the potential for sacred groves to contribute towards sustainable development in some locations is significant, although more research on the outcomes of projects is necessary. However, in addition to involving natural scientists in studying the biodiversity in sacred groves and at other sacred sites, and assessing their significance to surrounding ecosystems, it is absolutely crucial to enlist the services of anthropologists. These are needed in order to better understand the significance of these spaces within the religious and cultural traditions of local communities, in order to avoid adopting essentialist and romantic discourses about their significance and function. Moreover, different studies have pointed towards economic as well as non-economic functions of these groves. Deb (2007), for instance, in a study looking at sacred groves in West Bengal in India, argues that economic benefits include the availability for plants used in traditional medicine, the maintenance of safe drinking-water sources, the possibility of some biomass extraction (e.g. of wood for cremation) and ecological benefits for the surrounding forests and land. He argues that the role that sacred groves play in local systems of forest management could usefully inform 'new and sustainable' forest management regimes (ibid.: 30).

Summing up

This aim of this chapter has been to address the following questions: *Is it correct to assume that religious traditions teach that one should care for the natural environment? Is this question relevant in practice? In what ways might religion contribute towards 'sustainable development'?* Environmentalism is a modern phenomenon, responding to particular contemporary challenges that were largely unknown in the past. The global nature of ecological problems today, as well as the scale and potential longevity of their impact, brought about by industrial development and the overuse of natural resources, have radically altered the way that recent generations relate to the natural world and view its significance. This includes the emergence of concerted and often impassioned responses on the part of governments, NGOs and individuals to attempt to find ways of slowing down or reversing the negative effects of environmental destruction and climate change. Communities in the past may have acted to preserve natural resources in their local area or to protect areas of great aesthetic value, and religion may have played a role in these actions. However, there was no environmental movement and no sense that humans had the capacity to radically alter the future of the planet in such drastic and far-reaching ways.

Religious environmentalist discourses should be seen as modern-day interpretations of religious traditions, rather than straightforward reflections of the actual character of religion in the past. Just because religions can be interpreted to support modern environmental thinking, that does not mean that people who are members of those traditions think and would ideally behave like environmentalists, either today or in the past. Discourses about religions and the environment often draw upon romanticized images of an 'eco-golden age' that was sustained partly through religious or indigenous traditions, but are unlikely to reflect the way that people actually lived or thought about the significance of their religious traditions. As we have seen in this chapter, discourses about religions and the environment can be problematic when they are interpreted in a romantic way to imply that poor people are inherent environmentalists because of their religious traditions, or that they would necessarily choose to be if it were not for the encroachment of modern western modes of production. Moreover, the idea that 'putting the Earth first' is a post-materialist luxury that the poor cannot afford has also been suggested as a limitation of religious environmentalist discourse. However, such essentialist discourses are sometimes used strategically to shape and affirm shared identities within oppositional movements. It is also the case that although religious environmentalist discourses can be essentialist and romantic, projects that aim to harness ecological religion can have pragmatic goals that are less concerned with fundamentally transforming the relationship between humanity and the natural world (in line with some version of deep ecological or eco-centric thinking) and instead are more focused on achieving particular pragmatic ends, typically within so-called 'empty belly environmentalism' (Guha and Martínez-Alier 1997; Tomalin 2009c).

Particularly since the 'turn to religion', religious environmentalist discourses have been increasingly recognized by international organizations as relevant to the pursuit of different aspects of sustainable development, from biodiversity conservation and managing climate change, to the lobbying of governments and industry to adopt more sustainable policies and practices, as well as appealing to individuals to change their behaviour. There are many examples of initiatives in developing contexts that aim to harness 'ecological religion' to promote sustainable development, and a number of these are now supported by international organizations such as the World Bank or UNESCO. However, this type of engagement needs to be accompanied by the emergence of a well-developed academic or policy literature that examines the outcomes of such projects, the extent to which they could be usefully replicated elsewhere, the lessons learned and the places where caution needs to be exercised, with respect to securing sustainable development goals that will improve people's well-being and livelihoods. In contrast to studies that employ rather romanticized approaches to religions and the environment, such research is thin on the ground.

As we have seen above, in the case of environmentally sustainable development, international donors are increasingly recognizing the role of 'faith-based organizations' (FBOs) in different contexts. For instance, we discussed some aspects of the engagement between the World Bank and the UNDP with the Alliance of Religions and Conservation, which itself supports a broad range of environmentalist FBOs across the globe. In the following chapter we shall look more closely at the nature and activities of faith-based organizations working in development.

Questions for discussion

1 What are the main arguments put forward by religious environmentalists?

2 Assess the claim that religious environmentalism is an expression of the 'myth of primitive ecological wisdom'.

3 In what ways and with what effects did ecofeminism influence the 'women environment development' (WED) approach adopted by mainstream development actors during the 1980s and 1990s?

4 To what extent can religious environmentalism contribute towards 'sustainable development'? Or is it a romantic and post-materialist construction? In your discussion, draw upon the examples discussed towards the end of the chapter.

Recommended reading

Chapple, Christopher Key and Mary Evelyn Tucker (eds) (2000) *Hinduism and Ecology: The Intersection of Earth, Sky, and Water*, Cambridge, MA: Harvard Divinity School, Center for the Study of World Religions/Harvard University Press.

Foltz, Richard, Azizan Baharuddin, and Frederick M. Denny (eds) (2004) *Islam and Ecology: A Bestowed Trust*, Cambridge, MA: Harvard Divinity School, Center for the Study of World Religions/Harvard University Press.

Hessel, Dieter T. and Rosemary Radford Ruether (eds) (2000) *Christianity and Ecology: Seeking the Well-being of Earth and Humans*, Cambridge, MA: Harvard Divinity School, Center for the Study of World Religions/Harvard University Press.

Tucker, Mary Evelyn and Duncan Ryuken Williams (eds) (1997) *Buddhism and Ecology: The Interconnectedness of Dharma and Deeds*, Cambridge, MA: Harvard Divinity School, Center for the Study of World Religions/Harvard University Press.

The above edited volumes are some of the outcomes of ten conferences held at Harvard University around the theme of religions and the environment.

Palmer, Martin and Victoria Finlay (2003) *Faith in Conservation: New Approaches to Religion and the Environment*, Washington, DC: World Bank.

Whitten, Tony and Bryony Morgan (eds) (2006) *Faiths and the Environment*, Washington, DC: World Bank.

These two volumes are joint publications between the Alliance of Religions and Conservation and the World Bank.

Websites

http://fore.research.yale.edu/ This is the website of the Forum in Religion and Ecology at Yale. It is an excellent resource, providing outlines of the responses to ecological problems from different religious traditions, bibliographies, details of projects and resources for teachers.

www.arcworld.org The website of the Alliance of Religions and Conservations includes details of the projects it supports as well as information about its links with multilaterals, such as the World Bank and UN bodies.

8 Researching and understanding the role of 'faith-based organizations' (FBOs) in development

This chapter will:

- address the following question: *Do 'faith-based' development organizations have a comparative advantage over secular organizations?*;
- present a brief discussion of different types of religious organizations involved in welfare, humanitarian and development activities, with a focus on Christianity, Islam and Hinduism;
- examine the genesis of the term 'faith-based organization' and its entry into development discourse and practice;
- investigate the reasons why interest in FBOs within mainstream development policy and practice has increased over the past decade and why the number of FBOs seems to have grown;
- assess attempts to define FBOs and classify them in terms of typologies;
- outline and critically examine discourses about the 'distinctiveness' and 'comparative advantage' of FBOs that are currently popular within mainstream development policy and practice.

Introduction

Over the past decade, development donors have increasingly chosen to support the work of 'faith-based organizations' (FBOs), although such organizations have been involved in various types of charitable, philanthropic, humanitarian and development work for far longer. Many have also been funded by multi- and bilateral donors for years but, because of such donors' explicit secularism, and dislike of evangelism in particular, they have tended to downplay their religious identity and motives (especially the case in the secular European countries). An examination of FBOs in development provides very clear evidence of the recent 'turn to religion', with donors increasing funding to such organizations and in some instances choosing to favour them over 'secular' organizations. While the United States has gone furthest in doing so, a recent White Paper in the United Kingdom promised to double funding to FBOs (DFID 2009: 5). Rowan Williams, the then Archbishop of Canterbury, who hosted two inter-faith consultations in the run-up to the White Paper, responded positively to this promise, saying, 'I welcome the Government's clear commitment to engage with the *distinctive role of faith groups* as they maintain their long-held mandate to provide direct assistance to the world's poorest, while tackling issues of justice, peace and governance' (emphasis added).[1] More recently still, DFID has launched a new 'Principles Paper' (2012) to guide its collaboration with faith groups, underscoring the key principles of the partnership between faith groups and DFID as transparency, mutual respect and understanding.[2]

An understanding of faith groups or 'faith-based organizations' as having a distinctive role, or even a comparative advantage over secular organizations, has come to shape much discussion about the role of FBOs in development. The quotation from Clarke, presented in Box 8.1, provides an example of this discourse, which views FBOs as 'the forgotten factor' (Selinger 2004), marginalized from mainstream development owing to the secular and economistic focus of the western development agenda, but which have been recently (re)discovered and their unique contribution recognized (Tyndale 2006; Marshall and Marsh 2003). In a socio-political climate globally that can no longer ignore religions, for good or for bad, the 'faith-based organization' has acquired a significance that would have been difficult to predict even twenty years ago. Although many have welcomed the increased focus on FBOs as an attempt to bring a hitherto neglected body of stakeholders into the development process, others are more cautious (Stambach 2005; Balchin 2011). They perceive that there has been rather a rush to engage with FBOs without appropriate assessment of their effectiveness in achieving development goals, for instance those relating to gender (see Chapter 6), and argue that this a product of a broader political culture generated in the United States that (over-)values the faith contribution to welfare and social services.

Box 8.1

The distinctiveness of faith-based organizations (FBOs)

Clarke, for instance, characterizes FBOs as having

> a number of characteristics that distinguish them from their secular peers. They draw on elaborate spiritual and moral values that represent an important and distinct adjunct to secular development discourse. As a result, they have a significant ability to mobilise adherents otherwise estranged by secular development discourse. They are highly networked both nationally and internationally and are highly embedded in political contests and in processes of governance in both horizontal and vertical terms. They are less dependent on donor funding and they have well-developed capacity and expertise in the key areas of development practice. As such, they are important actors in the development process and warrant commensurate attention in development policy.
>
> (2006: 845)

This chapter opens with a brief discussion of different types of religious organizations involved in welfare, humanitarian and development activities, in Christianity, Islam and Hinduism. Despite decades of overlooking these activities, mainstream development actors are increasingly likely to view so-called faith-based organizations (FBOs) as important, and to encourage engagement between religious actors and development organizations, both locally and internationally. Next we will examine the entry of the term 'faith-based organization' into development discourse and policy, and the extent to which it first became popularized in reference to organizations that became prominent as a result of the US-led 'global faith agenda', which has been spurred by the religious right but is also a reflection of concerns over the rise of religious identities after 9/11.

The chapter then addresses the ways in which the term has been defined, because if a particular understanding of what an FBO is dominates within development circles, this may mean that other types of organizations are sidelined, with the result that they are less likely to be identified as potential partners, or consciously work to become more like FBOs in order to benefit from donor support. Reviewing definitions and typologies in current use, the chapter asks what exactly they mean by 'faith-based', whether they capture the diverse organizational forms that religious development activity in the Global South takes, and whether there really is a category of FBOs that contrasts in key ways with organizations that are 'secular'. It argues that a more nuanced understanding is needed of what constitutes an FBO, as well as a more complex account of what 'faith-based' actually means, noting that some question the viability of the term and concept itself.

As I have already noted, it is often asserted that FBOs are 'distinctive' (that is, they have characteristics that make them stand out from others) and even, more ambitiously, that they have 'comparative advantages' (that is, the distinct characteristics of FBOs place them in a better position than secular NGOs to engage in 'successful development'). The final section of the chapter will examine these debates, reviewing attempts to assess the claims, identifying some of the challenges of doing so and assessing the extent to which the debate reflects modern western discourses.

Religious organizations and 'development': from charities to FBOs

Different religious traditions make claims on individuals to carry out activities to help those less fortunate than themselves (e.g. the giving of *zakat* in Islam). Moreover, in all corners of the globe religion has a long association with various types of welfare and philanthropy that pre-date the emergence of development as a global concern, as well as the formal rise of the FBO in the 2000s (Tomalin and Leurs 2011). As Box 8.2 illustrates, today these 'faith-based' initiatives vary from those based in local churches or mosques, for instance, to charitable organizations established for the purpose of distributing food and shelter materials following disasters, or providing education and health care (Plates 8.1 and 8.2). Some of these organizations are informal (i.e. not registered under charity or NGO legislation), whereas others have been formally established and may operate locally and/or be branches of larger organizations based in the West or the Arab world in particular. Many of these organizations are linked to institutions that are centuries old and others can be traced to the colonial period. However, many more have emerged as the post-Second World War development project gained pace in the post-Bretton Woods era, particularly since the structural adjustment programmes of the 1980s prioritized NGO-led development. This has given rise to new pressures on organizations that wish to tap into the aid-business funding, and such organizations are required to meet certain criteria to be eligible for funds. By contrast, there is a wide range of organizations globally that raise their own funds, often from religious sources, and are not directly beholden to the restrictions of the aid business.

Plate 8.1 *Wat Thamkrabok, Saraburi, Thailand: a popular detox centre*

Source: Getty Images.

Detox residents are pictured cleaning up the temple grounds. Wat Thamkrabok runs a drug detox programme, for both Thais and foreigners.

Plate 8.2 *People attend an AIDS church in Nairobi, 25 March 2001*

Source: Getty Images.

The controversial Pentecostalist pastor John Nduati is pictured standing next to an 'overwhelmed' woman, who came to his God's Power Church to be 'saved' following abuse by her husband. Believers attend up to four times a week, hoping to be cured of AIDS, but also cancer, other diseases and addictions. In particular, most people attending the church will be too poor to afford medication to manage their illness.

Box 8.2

Types of religious organization involved in welfare, humanitarian and development activities in Christianity, Islam and Hinduism

Christian charities and FBOs

Many Christian welfare, humanitarian and development organizations have their roots in the colonial era. For instance, Christian missionaries were important providers of education, since without literacy people were unable to read the Bible. While this suggests that this emphasis on education was as much part of attempts to convert as to improve people's lives, there was a desire to help the poor, and missionaries also established hospitals, dispensaries, orphanages, hostels, infirmaries and seminaries, many of which have continued to operate since independence. Today there are different sorts of international Christian development FBOs: those that operate more like secular development agencies ('Oxfam with prayers' – Thomas 2005), such as Christian Aid, and those that are explicitly Evangelical in orientation, such as World Vision, Tearfund and Samaritan's Purse (see Box 9.1 on p. 235). While many of these organizations were initially set up for charitable purposes, over the past decade or so they have also begun to emphasize that they are concerned with 'development' as a long-term and sustainable process of change rather than just charity, which can create dependency and fail to bring about change. In addition to international Christian FBOs, there are many more local church-based organizations across the globe that may have limited links at the international and even national levels. These may be local churches or informal organizations linked to churches, often continuing long-standing traditions of support and service provision.

Muslim charities and FBOs

There are a large number of Muslim charities and FBOs globally involved in welfare, humanitarian and development activity, in both the West and the Muslim world. Places of worship, including mosques and Sufi shrines, also have an important function in this respect, in giving food to the poor or offering relief during and following disasters. Today there are many international Muslim FBOs that resemble formal NGOs (such as Islamic Relief or Muslim Hands), which also think about their work as contributing towards development rather than charitable ends alone, as well as smaller local Muslim charities, which are less likely to think about their work as 'faith-based' and indeed less likely to use the label FBO (Kirmani and Zaidi 2010). Moreover, Muslim charities and FBOs tend to focus their development, aid and relief activities within Muslim communities. Since 9/11, many of these organizations have been treated with suspicion and, as Kroessin writes, 'in the public mind, Islamic charity organisations have become little more than funding fronts for terrorism and jihad' (2007; see Burr and Collins 2006 for examples of this). However, in many cases investigations have failed to produce much evidence 'linking agencies or their staff with terrorist activity' (Kroessin 2007).

Hindu charities and FBOs

Compared to Islam and Christianity, we do not find as many development organizations linked to the Hindu tradition that are international in reach, possibly because most Hindus live in India (85 per cent of the population are Hindu), whereas Muslims have large populations in many countries. Those that do exist include Hindu Aid (based in the United Kingdom), the Hindu American Seva Charities (based in the United States) and Sewa International (with bases in multiple countries). However, there are many thousands of small and large Hindu charities linked to places of worship (e.g. temples, ashrams or study/meditation centres) or operating as separate organizations, which are involved in welfare, humanitarian and development activities. Places of worship are also directly involved in such activities as giving food to the poor or handing out medicines. Many of the activities of these organizations have been carried out for centuries and pre-date modern western interest in international development and humanitarian assistance. By contrast, other organizations (e.g. the Hindu American Seva Charities) have been established in response to the 'turn to religion' in both domestic and international affairs in the United States and clearly locate themselves within the discourse and practices of international development. Other large Hindu organizations in India, such as the Vishwa Hindu Parishad (VHP) and the Rashtriya Swayamsevak Sangh (RSS), engage in charitable, disaster relief and development work, in addition to their other cultural, religious or political activities (see Box 4.8 on p. 101; Jaffrelot 2008). As with some Muslim organizations, there has been concern (and indeed some evidence) that Hindu nationalist organizations such as the VHP have used money raised for humanitarian work to fund 'terrorist' activities.

Mainstream international development donors and NGOs increasingly recognize the impact of this long-standing 'faith-based' sector in welfare, humanitarian and development activities as significant. With respect to health, for instance, a number of recent studies stress the importance of strengthening faith-based engagement in public health policy and practice, and call for a better understanding of the nature and the scale of religious contribution to health (Olivier and Wodon 2011; Cochrane *et al.* 2011). Major development actors, including the World Health Organization (WHO), have also taken up this call. As is outlined in Box 8.3, the WHO published a study in 2006, *Appreciating Assets: The Contribution of Religion to Universal Access in Africa* (to HIV/AIDS treatment, care and prevention), documenting the 'contribution made by religion and religious entities to the struggle for health and wellbeing in Zambia and Lesotho, in a context dominated by poverty, stressed public health systems and the HIV/AIDS pandemic' (2006: 1). The study suggests that public health initiatives in Lesotho and Zambia have much to gain from engaging with a range of 'faith-based' actors and initiatives, but stresses that it is important to have an appropriate understanding of the local religious culture in each case, rather than assuming that the same approach can be applied everywhere (ibid.: 125).

Box 8.3

Appreciating Assets: The Contribution of Religion to Universal Access in Africa

This study, carried out in Zambia and Lesotho between November 2005 and July 2006, involved an inquiry 'into the nature and potential contributions of religious entities to the struggle against HIV/AIDS, to universal access to treatment, care and prevention, and to health and wellbeing more broadly' (WHO 2006: 2). While since the advent of the HIV/AIDS epidemic, there have been 'tensions around religion and public health, especially when it comes to HIV/AIDS and matters to do with sexuality, condoms and stigma' (ibid.: 131) the study found these to be 'part of the past' and that 'the sheer human impact of the pandemic is drawing religious entities into new and significant contributions to health and wellbeing' (ibid.: 131).

Overall, the

findings make visible – and in many cases map for the first time – approximately 500 religious and partner organizations working in the area of HIV/AIDS, some 350 at the local level. *These groups, in particular the community congregations, support groups and intermediary bodies, are seldom seen by policy-makers and often remain unknown even to their formal religious structures.* Our findings suggest these assets could and should be more effectively mobilized and linked for scale up to universal access.

(WHO: 2006: 2; emphasis added)

The study found that

[i]n Lesotho, approximately 40% of national health service is estimated to be provided by 9 Christian hospitals and 75 health centers whose managing churches are members of the Christian Health Association of Lesotho (CHAL); in Zambia, the numbers are similar, with 30% of services provided by 30 hospitals and 66 rural health centers whose managing churches are members of the Churches Health Association of Zambia (CHAZ).

(ibid.: 7)

However,

[r]eligious entities have also been shown to have other critical, but less tangible assets, including extensive volunteer networks, expert lay professionals, congregations/churches/mosques/temples in even the most remote locations, and deep systems of community support, 'voice,' advocacy, trust, consolation, and hope.

(ibid.: 7)

In the remainder of this chapter we are going to look more closely at the rise of the FBO in development policy and practice. In particular, we will explore the suggestion not only that FBOs are being taken more seriously as significant development actors by mainstream development donors and organizations (and that this is generally positive), but also that in important ways the western development project is itself shaping the organizational culture of FBOs, which, as Olson notes, is 'a product of development and not just a network through which development happens' (2008: 394; see also Vanderwoerd 2004).

The rise and rise of the FBO: from the 'religious right' in the United States to a global phenomenon

It is no coincidence that the frequency with which the term 'faith-based organization' began to circulate coincided with a rise in awareness of the resurgence of religious activity in public life globally, particularly in the post-9/11 period. More specifically, its origins can be traced to the era of the Republican government of George W. Bush in the United States (2001–2009), when there was an increased desire to draw religious organizations into public life around the agendas of social cohesion and service delivery (Amirkhanyan *et al.* 2009;

Plate 8.3 *President George W. Bush meets with religious leaders in the Roosevelt Room of the White House, 20 September 2001*

Gibelman and Gelman 2002; Johnson *et al.* 2002; Owens 2007; Wuthnow 2004; Plate 8.3). This favouring of religious organizations in the public sphere was matched by the availability of designated funding streams for both domestic and international projects and was a trajectory that influenced the priorities of the Labour government of Tony Blair in the United Kingdom (1997–2007) (Dinham *et al.* 2009; Torry 2005). The term 'faith-based organization' was used as a label for organizations that arose or reshaped themselves in response to the new political climate, which sought to elevate the role that faith traditions can play in many aspects of public life, including international development. While many already existed and engaged in development activities, often, new organizations have formed to engage in policy dialogue or gain access to donor or government funding. Box 8.4 outlines a series of factors, which, according to Clarke, can account for the apparent proliferation of FBOs globally.

Box 8.4

Factors that account for the proliferation of FBOs globally

1 The rise of the 'Christian right' in the United States following the election of Ronald Reagan as US president in 1980: this refers to a shift from the dominance of traditional Christian denominations towards Evangelical and Pentecostal adherence representing 'a more fervent, and ideologically right-wing, form of faith . . . [that has been] influential in the passage of legislation that guides US foreign policy' (Clarke 2008: 19).

2 Structural adjustment and the rise of NGO in the 1980s: in the United States, legislation was passed to outlaw discrimination against FBOs in the award of government funding (the 'Charitable Choice' provisions of the 1996 Welfare Reform Act and the 2001 Faith-Based and Community Initiatives Act), 'provoking concerns about the blurring of church–state boundaries and potential discrimination in favour of FBOs' (ibid.: 20).

3 The rise of political Islam 'provided a conservative counter-current to the US Christian right' (ibid.: 21). This included an increase in the number of Islamic FBOs, many funded from Saudi Arabia.

4 The decline of communism and the end of the Cold War stimulated a rise of 'identity politics', including movements and organizations reflecting religious identities.

5 The emergence of transnational civil society over the past two decades, including examples such as the Jubilee 2000 campaign (see Chapter 3).

6 Immigration to western nations from other regions of the globe has led to increased importance of multicultural and multi-faith dynamics. This includes the maintenance of links to countries of origin, a feature of which is support for charities that provide humanitarian and development assistance, as well as the substantial economic flows from remittances.

Box 8.5

International organizations and FBOs

The World Bank and the World Faiths Development Dialogue

Clarke and Jennings write that much of the impetus for the recent engagement with FBOs 'stems from the work of former World Bank President James Wolfensohn and former Archbishop of Canterbury Dr George Carey, and a series of conferences of donor representatives and faith leaders (in London in 1998, Washington, DC in 1999, and Canterbury (England) in 2002)' (2008a: 2). Wolfensohn and Carey established the World Faiths Development Dialogue (WFDD) in 2000 to respond to 'the opportunities and concerns of many faith leaders and development practitioners who saw untapped potential for partnerships'.[3] Further meetings took place in Dublin in 2005 and Accra in 2009. These initiatives were initially part of the World Bank's Development Dialogue of Values and Ethics (DDVE) and have resulted in several publications (Marshall and Marsh 2003;

Marshall and Keough 2004). The WFDD is currently based in Washington, DC, at Georgetown University and is involved in a number of ongoing projects, including a 'Global Mapping of Faith-Inspired Organizations and Development'.[4]

The United Nations Population Fund (UNFPA)

The UNFPA has decades of experience working with FBOs and has recently produced 'Guidelines for Engaging Faith-Based Organisations as Cultural Agents of Change' (UNFPA 2009). Beyond this, it has a number of publications that explore the role of religion and culture in its work (2005, 2007, 2008). The organization has been actively engaging with FBOs for far longer than most other international agencies and it seeks to build bridges between FBOs and secular development practitioners:

> Faith-based organizations, sometimes referred to as FBOs, also have a long history of making a difference in people's lives by delivering crucial social services. Faith-based service delivery networks continue to identify and reach out to those in need, often during the most difficult times and in the most remote areas. FBOs were among the earliest responders to mobilize around – and continue to be major contributors – to the worldwide response to AIDS.[5]

USAID

The USAID Faith-Based and Community Initiatives

> aim to better equip faith-based and [secular] community development organizations around the world as they provide assistance to people in need around the world. We do this by serving as a bridge between small non-governmental organizations and USAID; by connecting groups with relevant points of contact at USAID and informing groups of various partnership opportunities on a range of development challenges; and by encouraging collaboration among the people and organizations addressing human suffering around the world.[6]

Until recently, western donors and governments have been distinctly wary about engaging with and funding FBOs, particularly those that are Evangelical, or perceived as 'extreme' or 'fundamentalist', largely viewing 'religion as a negative force' (James 2011: 110). However, according to Clarke and Jennings, 'in little over a decade . . . there has been a gradual movement from estrangement to engagement' (2008a: 2). Both bi- and multilateral donors are increasingly choosing to fund and work with FBOs, and, as Box 8.5 illustrates, they have produced positive statements about the contribution of FBOs to development and poverty reduction (see also Box 1.4 on p. 9). In addition to increased support from donors, organizations and networks have been established with the aim of promoting and linking together the work of different FBOs in development with one another and with the broader development community (e.g. the Women, Faith and Development Alliance and Religions for Peace).[7]

Berger (2003) draws attention to the emergence of 'religious NGOs' on the global stage as having taken place through a series of international campaigns, involving both secular and religious actors. These included 'heated debates' between secular and religious organizations at the 1994 UN Conference on Population and Development, concerned with issues around family planning and the empowerment of women, but also campaigns in which religious and non-religious organizations have worked together on the same platform, such as the Jubilee 2000 debt cancellation campaign, the International Campaign to Ban Landmines and the advancement of the International Criminal Court (ibid.: 20). Many also consider that the Millennium Declaration (2000) and the MDGs lie at the heart of this new engagement with religions and faith, with 'faith communities and organizations to which they give rise . . . seen as important actors in galvanizing the moral commitment on which the MDGs depend and in popularizing them in local churches, mosques and synagogues' (Clarke and Jennings 2008a: 2; see also Marshall and Keough 2004: 4).

Despite the global rise of the FBO and the increased profile of FBOs in development, as well as the important role that many of these organizations play, there are a number of problems with the term and concept, particularly when it is transported to developing contexts. In the following section, the ways in which the term 'faith-based organization' has been defined, as well as attempts to develop typologies to aid our understanding, are examined.

Definitions and typologies

A famous Supreme Court judge once commented that he could not define pornography, but he knew it when he saw it. There are those who make the same kind of claims about what have come to be called faith-based organizations (FBOs).

(Jeavons 2004: 140)

Even in the United States (where the term 'faith-based organization' is most heavily used), commentators have expressed concern that it is poorly understood and defined, and that public debate about the 'purposes, roles and operations, and funding of these organizations' is inhibited because of the 'ambiguity and confusion of the terms of concepts' (ibid.: 140). (It has proved similarly difficult to define what an NGO is, and portrayals of 'NGO' as a homogeneous category have drawn criticism.)

In response, since the late 1990s there has been a steady stream of public policy-linked scholarship from the United States that has aimed to define what an FBO is and to elaborate on the different 'types' of FBO and how 'faith' is manifest in them. Smith and Sosin (2001), for example, argue that the description 'faith-based' excludes many organizations whose faith basis is implicit rather than explicit and suggest that the existence of a category of organizations with limited links to the secular world is not supported by their research. Instead, they prefer the term 'faith-related', recognizing that faith is manifest in different ways in organizations in different places, and for different reasons. They also note that 'agencies most closely aligned with faith must act in the secular world, while agencies with a strongly secular orientation might be influenced to some degree by their religiosities' (ibid.: 653). Sider and Unruh similarly argue that 'whether an organization is faith-based cannot be answered with a simple yes or no. The faith nature of organizations is multidimensional, requiring a range of types' (2004: 116). In response to this concern, they developed a framework that classifies organizations in terms of the degree to which faith is manifest in their work, along a spectrum from 'faith permeated' to 'secular'.

This body of literature from the United States emerged against the backdrop of a political culture that increasingly encouraged the involvement of FBOs in social welfare provision and introduced legal changes to make this possible. In response, studies sought to define and classify a range of predominantly Christian organizations that could potentially seek state funding while maintaining the constitutional church–state separation. Jeavons, for instance, argues that 'congregations should not be described as FBOs under any circumstances' (2004: 144). To partner with congregations or churches, or even separately incorporated organizations (i.e. distinct FBOs linked to a congregation) that are sectarian, it was argued, would present a threat to the church–state distinction.

Building on this work from the United States, but diverging from it in important ways, various studies have asked how we should define, classify and assess the potentially distinctive contribution of FBOs working in development. While some authors use the term 'faith-based organization' as if it describes a clear organizational category, and some organizations are content to call themselves FBOs, others problematize the concept of an FBO, arguing that the way the term is understood by donors reflects its genesis in a North American context and that alternative definitions and typologies are needed to understand it in developing contexts. For instance, Jeavons' (2004) argument that congregations should not be included within the definition of an FBO is linked to issues around the relationship between religions and the state that may not be relevant elsewhere. Thus, as Clarke and Jennings ask with respect to developing contexts:

> What exactly can, or cannot, be labelled as a faith-based organization? Do we, for instance, exclude Sunday schools, basic Christian Communities, or informal temple committees? . . . We use the term 'faith-based organization' in reference to any organization that derives inspiration and guidance for its activities from the teaching and principles of the faith or from a particular interpretation or school of thought within the faith.
>
> (2008a: 6)

In Clarke's view (2008), to limit the definition of FBOs to formally registered organizations that resemble NGOs would exclude much religiously inspired development work with which donors might usefully engage. This wider set of organizations includes places of worship and congregations, apex bodies representing religious hierarchies, missionary organizations and religiously based socio-political groups (such as political parties). For instance, religious places of worship and their congregations are often involved in various forms of development work, sometimes through other organizations (e.g. Catholic parishes raising money for Caritas), sometimes directly (e.g. digging wells, giving food to the poor) and sometimes as 'partners' of overseas organizations and funders. In addition, to distinguish NGO-like FBOs from religious organizations that have been doing philanthropic work and service delivery for hundreds of years in some countries and since the colonial period in others is both artificial and arbitrary.

In addition to the issue of what to include in the definition of FBO, the very act of distinguishing between 'faith-based' and secular organizations in contexts where religion may permeate almost all aspects of life is arguably an imposition of Christianized or western understandings of the relationship between the religious and the secular that do not have the same meaning elsewhere. Not only are many members of so-called secular organizations motivated by religious values, but organizations operate in contexts in which the secular and the religious are not clearly differentiated. This issue of the extent to which it is possible to separate the religious from the secular in practice may render the term 'faith-based organization' fairly meaningless in many contexts, where it may not be a term used to refer to humanitarian organizations, even those with an apparent religious basis. For example, in Pakistan, Islamic charities in Karachi do not identify themselves as 'FBOs' (Kirmani and Zaidi 2010; Kirmani 2012). Another reason why the term 'faith-based organization' is not used is that religion is often a sensitive issue, leading many organizations to prefer not to identify themselves as 'religious' or 'faith-based'. For example, some apparent FBOs do not use the label because they feel that it would restrict their capacity to obtain international funding (e.g. in Tanzania; see Leurs et al. 2011). Finally, the term itself is sometimes considered problematic because of its western origins and explicitly Christian overtones (e.g. the notion of 'faith', which is foreign to religions such as Hinduism and Buddhism). Thus, there is much else that is missed by FBO terminology – because conventional understandings of the term are too narrowly defined, people do not like it for cultural or political reasons, or they simply do not think of what they do in terms of its being 'faith-based' or not.

In order to avoid essentializing organizations as either 'faith-based' or 'secular', several authors prefer to talk about the 'pervasiveness' of religion in their aims and operations (Berger 2003). The extent to which faith is manifest in an organization is of interest to donors, some of which 'would like a sanitised separation between the institutional and spiritual elements' (James 2009: 10), leading them to favour FBOs that are not explicit about their religiosity and do not consider conversion to be a feature of their development activity. Grills (2009), for example, writes that multilateral organizations' increased interest in engaging with FBOs presents challenges to their Enlightenment ideology, with the result that they have sometimes imposed a condition that a recipient FBO removes faith activities from its programmes. While this separation of overt religious expression from development activities has been a concern of donors from countries with a history of constitutional stipulation of religion–state separation, in the context of international development, where such secularism is either different or inapplicable, its suitability needs to be questioned, since it may not reflect local relationships between religions, the state and society. In the context of donor–FBO relationships in developing contexts, this question of the extent to which degree of religiosity should be factored into decisions about specific partnerships needs to be revisited.

Despite these problems with the term 'faith-based organization', it is nonetheless increasingly widely used. However, the great variety of organizations potentially captured by the label is increasingly recognized. Thus, while international FBOs are more likely to resemble the 'western' model of an FBO that resembles an NGO, many local organizations are less likely to fit this mould (Kirmani and Zaidi 2010), leading to attempts to develop more complex typologies than the simple 'faith-based'/secular divide.

Typologies serve a range of different functions and can be used for different purposes. For instance, they can be used to identify distinct types of organization, classify a large set of organizations or locate an individual organization in a universe of different organizational types. They can assist researchers by identifying the possible existence and characteristics of different types of organization in a particular setting and helping with the selection of new organizations to study; or they can inform policy makers about the types of organizations that exist and the potential opportunities for engagement with them. Hefferan *et al.* (2009) suggest that using a typology developed in one context as a starting point for understanding another can be useful, making revisions as necessary to take account of different circumstances. Such a strategy can raise interesting questions about the nature and role of 'faith-based' activity in different places (Tomalin and Leurs 2011). While typologies are never comprehensive and may obscure as much as they reveal, they are nonetheless useful in extending descriptive definitions.

It is important to be aware of the audience to which a typology is directed. For instance, the majority of current FBO typologies reflect the US context and were designed to speak to a policy audience in the United States, particularly one that was faced with increased opportunities, as well as mounting political pressure, to engage with 'faith-based' groups for domestic purposes. This literature is often taken as a starting point for developing typologies that have relevance elsewhere, but there have also been attempts to produce typologies and frameworks for capturing the range of FBOs engaged in different sorts of development work internationally. In the remainder of this section, two of these will be discussed.

Clarke (2008) suggests that there are five types of FBO relevant to international development: representative organizations or apex bodies; charitable or development organizations; socio-political organizations; missionary organizations;

and radical, illegal or terrorist organizations. He argues that donors have typically engaged with FBOs of the second type, but that these are normally Christian. In addition, he suggests that mainstream development agencies have shown a preference for working with FBOs that express their faith identity in passive ways (ibid.: 33). His analysis leads him to urge that there is a need to expand our view of the different types of FBOs beyond those with which donors have typically engaged. Clarke developed his typology in the course of work commissioned by DFID and it has a clear aim and audience: it aims to speak to mainstream donors in terms of categories with which they are familiar (for example, 'radical, illegal or terrorist organizations' is a controversial label, yet speaks in a language encountered in government and donor circles) and highlights the types of organizations that donors are likely to encounter.

While Clarke's aim is to expand our view of the types of organization that exist within the broad category of FBO, Hefferan *et al.* (2009) adapt Sider and Unruh's (2004) approach, which categorizes FBOs in terms of the extent to which faith is manifest in different aspects of their work. Sider and Unruh suggest eight factors as a basis for categorizing an organization: its mission statement, founding, affiliation, controlling board, senior management, other staff and sources of support (financial and non-financial). At one end of the scale are 'faith-permeated' organizations, in which faith is manifest across all the dimensions of an organization and its work, and at the other are 'secular' organizations. Sider and Unruh argue that classifying FBOs in this way could help funders choose appropriate organizations with which to work and also help FBOs to understand and describe their religious character – 'for purposes of strategic planning, fundraising, and evaluation' (ibid.: 132).

Hefferan *et al.* (2009) adapt this typology to analyse the FBOs discussed in their edited volume. The main differences that they identify between the developing countries described in their book and the American context concern the professionalism and formal organization of FBOs. Unlike the organizations studied by Sider and Unruh (2004), most of which have formal written mission statements and bureaucracies with management structures, many of the organizations discussed in Hefferan *et al.*'s book are informal and small, with little formal bureaucracy (2009: 18). In addition, whereas Sider and Unruh's typology presupposes a context in which there is a split between church and state, FBOs may not work in such a context in developing countries. For instance, Sider and Unruh believe that 'faith-permeated' organizations are wary about forming partnerships with government, as these might compromise their religious ideals, whereas in developing contexts such partnerships may even enhance some FBOs' ability to evangelize (Hefferan *et al.* 2009: 18).

The above discussion has indicated that attempts to define and classify FBOs in different contexts are fraught with conceptual and methodological problems. From a conceptual perspective, there is no agreed-upon definition of what an FBO is, and the often taken-for-granted distinction between 'faith-based' and 'secular' is unhelpful in describing and understanding organized expressions of religiously inspired welfare and development activity. Indeed in many contexts, especially in developing countries, use of the description 'faith-based' is uncommon, or even deliberately avoided. Further, the concept maps rather weakly onto the diversity of religious organizations involved in development globally, arguably reflecting a western Christian NGO-like model of a 'faith-based organization', one that may be difficult to find elsewhere or at most captures only part of the picture.

This debate about how we define an FBO points towards the ways in which discourses about FBOs both shape and reflect dominant ideologies about and approaches to development, and is important in a number of ways. First, it has ramifications for which organizations might be funded by donors, with those with recognizable characteristics and agendas perceived to be relevant more likely to be funded. Second, the concept and term 'faith-based organization' is, to a large extent, a product of the NGO-ization of development, which is ideologically as well as practically driven, often by the international agencies, and influences who works together, potentially marginalizing other actors. Third, there is evidence that organizations are explicitly shaping their activities and discourses to reflect the emergence of the FBO and to benefit from new opportunities to engage with donors. For some organizations this means reframing their religious identities in ways that are more attractive to secular donors, including downplaying an explicit or evangelical faith expression. In practice, not only does this not resolve the unease in formal development circles about combining development with religious evangelism, but also organizations that are considered to be evangelical are more likely to live in the shadow of criticism about the extent to which they are first and foremost committed to development. Finally, and perhaps most critically, the debate about how we define an FBO draws attention to the way in which the transformatory potential of the recent turn to religion in development policy and practice is weakened and threatened by its reliance on a particular model of 'faith-based' engagement with development. While donor agencies claim their engagement with FBOs as evidence of their acceptance of the importance of incorporating cultural and religious values and practices into development policy and practice, the above analysis suggests instead that religion is being brought into development in ways that reflect not only western perceptions of the nature of development processes but also western understandings of and attitudes towards religion. The perceived bias towards organizations that do not explicitly articulate their faith identity and a model of religion that neatly separates the religious from the secular suggests that donors are attempting to manage and neatly contain religion in ways that do not correspond with the way it manifests itself in many developing-country contexts.

The emergence of a western-influenced and arguably donor-shaped model of an FBO has been accompanied by the perception that FBOs make a 'distinctive' (often better) contribution to development than secular organizations. The final section of this chapter reviews a number of analyses to evaluate the extent to which the evidence available supports the claims made for FBOs.

Assessing the role of FBOs in development: evidence and challenges

Berger, quoting from the World Conference on Religion and Peace website, describes the widely held perception that FBOs are

> without question, the largest and best-organized civil institutions in the world today, claiming the allegiance of billions of believers and bridging the divides of race, class and nationality. They are uniquely equipped to meet the challenges of our time: resolving conflicts, caring for the sick and needy, promoting peaceful co-existence among all peoples.
>
> (2003: 17)

It is important, therefore, to examine who is making these claims and why, as well as whether or not there is evidence for them. Are they 'assumptions' or do they have a more solid basis? Such perceptions, which assert not only that FBOs are 'distinctive' but often also that they have advantages over secular organizations, sometimes appear to be made with little regard for their validity and to be over-general and inflated.

It is unsurprising that religious organizations themselves make such claims, which reflect their belief in the contribution religion can make to 'holistic' development. Moreover, religious actors arguably feel the need to promote their work because until recently FBOs were perceived to have been marginalized by mainstream development actors. However, the views are also held by some governments, international agencies and donors. For instance, when George W. Bush created the Office of Faith-based and Community Initiatives in 2001, he argued that they were 'the best, the most efficient purveyors of social services . . . since they often worked with low overhead and volunteer labor' (Hefferan *et al.* 2009: 6; see also Chaves 2002; Chambre 2001), and James writes:

> [T]he donor context for faith is changing as donors recognise that *many FBOs, even more than NGOs*:
>
> - provide efficient development services
> - reach the poorest
> - are valued by the poorest
> - provide an alternative to a secular theory of development
> - ignite civil society advocacy
> - motivate action.
>
> (2009: 2; emphasis added).

In addition to religious actors and donors, the notion that religious organizations are 'distinctive' has increasingly been discussed in the academic literature. Studies that aim to demonstrate the distinctiveness and/or comparative advantage of FBOs, often on the basis of one or more case studies of individual organizations, seem to be the most common (e.g. Hoksbergen 2005; Hoksbergen and Ewert 2002; Marshall and Keough 2004; Marshall and Marsh 2003; Tripp

1999). Much of this literature appears to share the perceptions described above and is arguably biased towards a positive view of FBOs. For example, Hefferan *et al.* observe that 'frequently underlying these recent works is an assumption about and endorsement of the power and promise of "faith-based organizations" to make international development more efficient, effective, and relevant than it is currently' (2009: 6). However, some analyses are more critical. For example, the 'distinctive' character of FBOs may be identified as a 'problem', since it is suggested that it makes them less likely to deliver the 'type' of development that donors value. In addition, FBOs are often seen as problematic if they seek to recruit new adherents or if they are perceived to show a preference for helping members of their faith group above others (De Cordier 2009: 614). In addition, a few studies challenge the basic assumption that FBOs are distinctive and suggest that in some situations they operate much like secular organizations, despite their religious identity (e.g. in rural Tanzania; see Green *et al.* 2013).

In the remainder of this section, some of the available studies are reviewed to assess some of the evidence on whether or not the claims of FBO distinctiveness and comparative advantage can be sustained. A number of challenges face any attempt to assess the supposed distinctiveness or comparative advantages of FBOs. First, such an analysis needs to have a clear idea of what an FBO is in order to compare it with non-FBOs. However, as we have seen, not only is there no agreed-upon definition, but also it is not clear what 'faith-based' means, and the distinction between 'faith-based' and secular organizations is not always easy to draw. In addition, the discussion of the different ways in which faith may be manifest in organizations' aims and approaches suggests that general statements about the distinctiveness of 'FBOs' may be misleading if they treat 'FBO' as a homogeneous category.

Some studies of FBOs explicitly attempt to demonstrate the distinctiveness or comparative advantage of FBOs in particular settings (e.g. De Cordier 2009; Lipsky 2006), but others do not systematically compare FBOs with secular organizations or specifically address questions about distinctiveness or comparative advantage. Nevertheless, some of them draw attention to the ways that some FBOs promote alternatives to, or a broader view of development than, secular approaches (Bornstein 2005; Occhipinti 2005). However, it is important to recognize that challenges to dominant development paradigms are not just expressed by FBOs and that FBOs are not always radical in this sense: a wide range of 'secular' organizations have challenged the economic focus of main-stream development policy as well as the tendency to ignore broader approaches to well-being that are connected to the 'moral base' of communities. Moreover, some studies have suggested that when NGOs or FBOs scale up and receive government funds, 'their legitimacy as non-governmental actors is eroded and their relationship with clients at the field level is compromised' (Nishimuko 2008: 178). This is clearly a potential risk for FBOs as they become more embedded in the donor-driven development process. De Cordier, for instance, highlights this as a risk for western Muslim 'faith-based organizations' that may currently have a 'comparative advantage' over secular NGOs, since they are viewed by local populations as in tune with their values and needs. He suggests that if these FBOs are viewed as 'subcontractors' for western donors, they may well lose their advantage and attract a backlash (2009: 623).

Another challenge in assessing FBOs in development relates to whether it is possible to assert with any confidence that a presumed 'distinctive' attribute of an FBO is a product of its faith identity. Many qualities that supposedly distinguish FBOs are also characteristics of some secular organizations, suggesting that the distinction may not so much be between 'faith-based' and 'non-faith-based', but, for instance, between organizations that are local and embedded in a community and those that are more distant and formal. This should force us

---

Box 8.6

The role of 'faith-based organizations' in building the democratic process: achieving universal primary education in Sierra Leone

Nishimuko writes:

> Although the special attention given to FBOs by academia and western donors is a relatively recent phenomenon, religion has long been an important source of culture and social structures. Colonialism brought Africa formal education through missionaries, and Islam also established formal and non-formal education.
>
> (2008: 173)

Moreover, he argues that people in Sierra Leone trust religious leaders, particularly considering the role that they played in bringing the bloody civil war to an end, and that FBOs provide 'emotional, moral, and spiritual support to people and their significant influence can mobilize the communities' (ibid.:173). His study is interested in exploring the role that religions and FBOs are playing in Sierra Leone in building democracy through the provision of primary education. Following the civil war between 1991 and 2002, around 70 per cent of the population lives in poverty and '[a]lthough secular municipal schools are available, about 75 percent of primary schools are faith based and are organized according to either Christian or Islamic principles' (ibid.: 174). In addition, the government, while limited in its own capacity to extend primary education,

> asks FBOs to propose and implement projects, and offers grants for them to do so. This is because even in remote rural areas there are churches or mosques and networks of FBOs, which are very helpful in establishing schools and sensitizing communities about the importance of education.
>
> (ibid.: 177)

In particular, 'FBOs in Sierra Leone contribute to practical matters, including school construction and rehabilitation, provision of school materials, providing scholarships to pupils and teachers, offering in-service training for teachers and compensating some teachers for non-payment of salaries' (ibid.: 178).

to question why the distinction between faith-based and secular is the one that is employed and whether this emphasis on 'faith' indicates that other factors are at play. Related to this, it is in practice difficult to assess whether or not, and in what ways, 'faith' influences the nature of an organization and how it carries out its activities, since faith cannot easily be disentangled from other influences such as the impact of donor practices; the interplay of local, national and global political forces; historical, cultural and social factors; and the nature of the particular development 'problems' being addressed.

One often-stated advantage of FBOs is that they are more trusted than secular organizations. For instance, as Box 8.6 illustrates, Nishimuko (2008), in a study of the role of FBOs in providing education in post-conflict Sierra Leone, claims that his findings demonstrate the trust that people place in religious organizations and leaders, partly reflecting the important role that they played in bringing the civil war to an end. FBOs are the most important providers of education, with about a quarter of parents interviewed saying that they chose schools for their children for religious reasons. However, the claim that this is because FBOs have a comparative advantage, because they are trusted and have long-standing roots in communities, is somewhat weakened when it is recognized that other providers are largely absent. If parents were given the chance, it is possible that they might make other choices. This is not to say that 'faith' is not a factor, rather that it needs to be seen in conjunction with other influences and constraints on people's choice of service provider. While Nishimuko does not extrapolate from his findings to make more general claims about the added value of FBOs, there is a danger that studies which employ the language of comparative advantage may play into broader problematic discourses that essentialize a 'faith-based' contribution as both homogeneous and distinctive. Nonetheless, Nishimuko's study does illustrate the very significant role that FBOs are currently playing in Sierra Leone with respect to the provision of primary education as a key dimension in ensuring the success of democratic processes in the future.

While there are a growing number of studies that aim to demonstrate that FBOs have a distinctive contribution to make, it is difficult to compare their findings because they investigate the role of FBOs in different sectors, with respect to different criteria, in different contexts and often using different definitions of what an FBO is (Amirkhanyan *et al.* 2009). This suggests that while some studies may indicate that FBOs have comparative advantages which enable them to undertake selected development activities or operate in some contexts more effectively than other actors, to generalize from individual studies would be misleading. Some researchers have aimed to move beyond case studies of organizations to undertake more systematic comparisons. For example, Lipsky (2006: 26) writes that, despite the historic involvement of FBOs in the delivery of health services in sub-Saharan Africa, with a more recent focus on HIV/ AIDS, 'there is little comprehensive research on the advantages that FBOs bring to the delivery of health services'. From existing studies she attempts to assess the extent to which FBOs are worse, the same as, or better than secular NGOs engaged in health care with respect to various criteria. On the basis of the evidence available, she suggests that FBOs are more likely to score highly in terms of moral and ethical standing, understanding of the local context, flexibility, and the ability to mobilize energy and resources. However, in terms of transparency and accountability, providing feedback to donors and governments, building constituencies and helping communities form their own representative bodies, they tend to do less well than NGOs.

Lipsky comments that there have been remarkably few attempts to undertake this sort of analysis, possibly because of the conceptual and methodological problems associated with defining and identifying FBOs, but also because of the absence of a comprehensive theoretical framework for comparative analysis (2006: 26). She proposes a framework that compares FBOs and NGOs in terms of their organizational traits, roles, and beneficiary empowerment. Although her aim is to help to establish the benefits of partnering with FBOs, and her analysis draws on secondary data and is impressionistic, her framework could be employed in studies based on more detailed primary fieldwork.

This brief review of studies that purport to assess the distinctiveness and comparative advantages claimed for FBOs by themselves, their supporters and many academic analysts shows, first, that definitional and methodological challenges often undermine attempts at objective and systematic assessment, and are not always acknowledged by analysts. Not only have there been relatively few attempts to assemble evidence on the nature and outcomes of FBOs' activities, but many of the studies that are available are case studies of individual 'faith-based organizations' rather than comparative assessments of FBOs and other organizations. There is little evidence to support a generalized claim that FBOs are distinctive and have a comparative advantage over secular NGOs. Much of the literature, therefore, reflects the global trend towards favouring 'faith-based' contributions to development and the desire to bring faith to the fore following its earlier marginalization in mainstream development discourse and policy. In assessing claims about FBOs' distinctiveness, therefore, it is important to be attuned to the broader political context and the ways in which it has influenced FBOs' engagement in development, their portrayal of their role, analysts' points of view, and the status accorded to rigorous methodology and empirical evidence in research on the sector.

Summing up

This chapter has addressed the following question: *Do faith-based development organizations have a comparative advantage over secular organizations?* Although we ought to take seriously the argument that religious organizations have the potential to make important contributions to development, contributions that have certainly been overlooked by mainstream development actors in the past and are not always taken seriously today, it is also necessary to

recognize that the way in which the debate about the distinctiveness or comparative advantage of FBOs is currently articulated reflects political priorities
as well as a desire to promote FBOs as viable development partners, worthy of
donor support. An answer to the above question is complicated by the very fact
that there is no clear idea about what an FBO actually is. Furthermore, as a term
and concept it can be traced to a particular way of viewing the contribution of
religious organizations to service provision that has its genesis in the United
States. Whether the term is wholly useful in developing contexts in terms of its
ability to capture the broad range of organizational contributions from religious
traditions to service delivery, advocacy, welfare or disaster relief has been
questioned. Instead, in terms of current usage as well as the extent to which
religious organizations in developing contexts actually use the term, it would
seem to have the ability to capture only a rather narrow selection of activities.

The language employed suggests that there are clear differences between both
'faith-based' and secular organizations and faith and non-faith approaches to
development, and that what is distinctive about different contributions can be
clearly demonstrated and traced back to the influence (or absence) of 'faith'.
However, the distinction between 'faith-based' and secular organizations may
not be useful in contexts where religion permeates almost all aspects of people's
lives, the model of an NGO-like FBO does not capture all types of religious
organizations engaged in development activities, and the term 'faith-based' may
reflect a largely Christian view of religion and a western (mainly US) context
characterized by particular forms of secularism. In addition, the studies available
to date do not indicate exactly what contributions religious organizations have
made to development, so it is not possible to draw conclusions about whether
their contribution can be captured through the language of 'distinctiveness' or
'comparative advantage'. Moreover, there is little evidence that many of the
supposedly distinct characteristics of FBOs are exclusive to them, or more
prevalent in them than in other sorts of organizations.

This is not to suggest that FBOs do not have a 'comparative advantage' in some
situations, but rather to emphasize that generalized claims cannot be made.
Instead of being avoided (for example, because they are perceived to mix
evangelism with development) or embraced (because it is assumed that they
share a distinctive set of characteristics that afford them comparative advantages), a more context-sensitive and case-by-case approach is recommended.
Realistically, it is often too complex and time-consuming to assess the full range
of possible partner organizations that exist in a given context, but studies that
highlight situations or activities in which FBOs appear to be more effective and
have had greater impact, as well as analyses that indicate where they have
created problems, can assist development donors and practitioners to better
understand the potential contribution of FBOs to development in different
contexts.

Many of the studies reviewed in this chapter use the language of distinctiveness
or comparative advantage, but sometimes this reflects a bias in favour of FBOs,

which runs the risk of concealing both the potential contribution of other partners and the problems that may be associated with religious organizations. I would argue that the notions of distinctiveness and comparative advantage always need to be carefully contextualized and may be best avoided, unless the analysis is truly even-handed and open-minded.

There is undoubtedly a need to develop a wider knowledge of FBOs, given their previous neglect in development studies. In particular, studies are needed that aim to be objective with respect to the positive and negative outcomes of FBO and NGO characteristics and activities. Rather than attempting to generate a list of the distinctive qualities of FBOs or to assess their advantages compared to secular NGOs, it may well be more useful to focus on all types of non-state organization that are active in a particular sector, attempting to address a particular development issue, or are operating in a selected geographical area. Studies also need to take into account organizational characteristics other than religion, such as whether organizations are international or national, and large or small. Such approaches would enable better comparisons to be made and generate hypotheses for further testing. I contend in this chapter that donors' desire to know the answer to the question of whether FBOs have comparative advantages reflects their location within a global political discourse that has 'found faith'. However, it is argued that framing the question in this way has led to misleading and simplistic answers, so that it is arguably the wrong question.

Questions for discussion

1 What factors can be suggested to account for the apparent rise in the number of FBOs working in development?

2 To what extent is the term 'faith-based organization' useful? Discuss its benefits and drawbacks.

3 What are the main problems with current discourses about the comparative advantage or distinctiveness of FBOs working in development?

4 In what ways could current approaches to conceptualizing FBOs be improved in order to better account for the reality of religious engagement with development?

Recommended reading

Bornstein, Erica (2005) *The Spirit of Development: Protestant NGOs, Morality, and Economics in Zimbabwe*, **Stanford, CA: Stanford University Press.** This book explores religious NGOs as important sources of humanitarian aid in Zimbabwe, where the state is weak and is unable to provide basic services.

Clarke, Gerard and Michael Jennings (eds) (2008) *Development, Civil Society and Faith-Based Organizations: Bridging the Sacred and the Secular*, Basingstoke, UK: Palgrave Macmillan. This edited volume examines the phenomena of 'faith-based organizations' (FBOs) and the role that they increasingly play in managing aid, providing services such as health and education, defending human rights and protecting democracy. It argues that greater engagement with FBOs is necessary to achieve, for instance, the Millennium Development Goals.

Hefferan, Tara, Julie Adkins and Laurie Occhipinti (eds) (2009) *Bridging the Gaps: Faith-Based Organizations, Neoliberalism, and Development in Latin America and the Caribbean*, Lanham, MD: Lexington Books. This book explores the ways in which 'faith-based non-governmental organizations' have proliferated as they attempt to solve the problems brought about by neo-liberalism. The chapters are grounded in empirical case studies and demonstrate the importance of ethnography for understanding the role of faith-based agencies.

Tyndale, Wendy (ed.) (2006) *Visions of Development: Faith-Based Initiatives*, Aldershot, UK: Ashgate. Tyndale's book is written from a background of decades of experience working in human rights and development. The chapters in the book focus directly on stories concerning the ways in which different groups and movements in Africa, Asia and Latin America perceive development as linked to religion.

Websites

http://berkleycenter.georgetown.edu/wfdd This is the website of the World Faiths Development Dialogue, which 'works to bring to light the efforts of faith-inspired individuals and organizations working to address local and global humanitarian challenges, and to bring together leaders from the worlds of religion and development to advance partnerships that have as their goal the elimination of global poverty'.

www.intrac.org/pages/en/faith-based-organisations-and-development.html INTRAC, a UK-based research, training and consultancy organization, which aims to support and strengthen civil society, has many years; experience with FBOs. Its website has a series of publications examining the particular advantages and challenges facing FBOs in development.

www.religionsanddevelopment.org The working papers in the 'publications' section of the website of the 'Religions and Development' (RaD) research programme has sections on papers in the following categories: 'faith-based service providers and their relationships with the state'; 'the development activities of NGOs and FBOs'; and 'mapping the activities of FBOs in development'.

9 Conclusion: religions and international development in the twenty-first century

This chapter will:

- address the following question: *What are the implications of the 'rise of religion' for mainstream development policy and practice in the twenty-first century?;*
- summarize the main factors that account for the recent 'turn to religion' by international development actors;
- assess the nature of the engagement between religions and mainstream development actors;
- examine a series of considerations that are relevant to any attempt within mainstream development studies, policy and practice to engage with religion.

Introduction

One aim of this book has been to examine the reasons for the uneasy relationship that has existed between the post-Second World War donor-led development project and religious traditions. The secular and modernist leanings of western development studies, policy and practice have meant not only that religions have tended to be viewed as backward and irrational, but also that they were predicted to become less significant over time, at both the public and the private levels. Until recently there had been very little attempt to engage with the institutions and leaders of religious traditions, the organizations that they generate and the ways in which religious beliefs shape people's worldviews, having a bearing on their relationships to economic systems, how they organize their family lives and the ways in which they deal with the stresses brought about by poverty and inequality. The previous chapters have highlighted the ways in which religious traditions in developing contexts continue to penetrate many aspects of social and political life. This contrasts with the reality of most modern western secular societies, which, if not completely secularized, certainly relegate religion to specific roles where it has far less impact than it did in the past. Moreover, as Taylor (2007) points out, in modern western societies secularity also carries with it the possibility of 'unbelief' where people can choose not to be religious. In most cultures throughout history, unbelief has not been an option, and in many that continues to be the case to this day.

However, more recently – roughly since the start of the twenty-first century – a number of commentators have noted that development donors and agencies are increasingly supporting and forming partnerships with different sorts of religious organizations, are inviting the views of religious leaders on development concerns and are undertaking or commissioning research looking at the relationships between religions and development. In Chapter 2 we explored the evolution of approaches to international development such that considerations of religion and culture are more likely to enter mainstream international development agendas. These were captured by Thomas (2005: 223–224) in terms of five shifts:

1 There has been a reassessment within the international aid business of the emphasis placed on economic growth and industrialization.
2 International development actors have begun to recognize the costs of disregarding people's values in their pursuit of understandings of development and modernization modelled on the trajectory of countries in the Global North.
3 There is now less inclination for mainstream development actors to see development in 'positivistic' terms.
4 Culture is becoming part of the official donor-driven development agenda.
5 The larger 'crisis of modernity' means that the West is forced to question its assumptions about modernization and secularization.

While these capture many of the points that we have examined in previous chapters, do they give the impression that post-Second World War donor-led development today is rather more open to considering religion and culture than it actually is? Moreover, do they suggest that solutions to the apparent 'crisis of modernity', which require a re-evaluation of assumptions about modernization and secularization, are rather more advanced than they actually are? In this concluding chapter the aim is to address the following question: *What are the implications of the 'rise of religion' for mainstream development policy and practice in the twenty-first century?* We will assess the nature of the engagement between religions and international development, and highlight a series of issues that it is important to consider in rethinking the relationships between religions and international development.

Globalization, transnationalism and the 'resurgence of religion'

One of the key themes of this book has been an examination of the extent to which processes of modernization and globalization have not resulted in the disappearance of religion (Berger 1967, 2000; Martin 1979; Casanova 1994). Instead, a number of theorists have turned their attention to the ways in which globalization has allowed diverse religious traditions to flourish rather than dwindle, as it was presumed they would under the influence of the spread of western secular values. Evidence against secularization theories can be found, for instance, in the increased transnationalism of many religious traditions, brought about partly by migration from South to North and the emergence of culturally diverse diasporas, and also of religious organizations and movements, which are increasingly important players in global politics, international development and civil societies more broadly. Beyer, for instance, identifies a specific role that religions can play in modern societies across the globe, where many

> locate themselves largely in the gaps left by the more dominant systems [e.g. politics, the economy]. They found oppositional movements, they take critical stands towards the dominant systems, often locating the roots of local and global problems in the operation of these systems. They gravitate in their operation towards the marginalized people and regions of the world. They create opportunities for sectarian flight from the world of the dominant systems.
>
> (2006: 103)

He draws attention to two types of religious response to globalization, and particularly to the problems that it creates (Beyer 1994). These reflect different tendencies within globalizing processes, where globalization, which is both a product of and a condition of modernity, results in a 'shrinking world' where national and ethnic boundaries are often dissolved but also where particular identities are frequently hardened. First, 'liberal' or 'ecumenical' types of religion have tended to participate in the globalized, transnational social movements that stress liberal values. Beyer suggests that ecumenical or liberal religious politics include liberation theology in Latin America and modern forms of religious environmentalism (1994). The concerns of such movements are focused upon freedom, tolerance, individualism and other globalized values, such as environmental or human rights. A recent example of this is to be found in the growing religious presence within the 'Occupy' movement, the anti-Wall Street and anti-large-corporation movement that emerged in Canada in 2011 and rapidly spread to other nations across the globe, both developed and developing (Plate 9.1).

Second, according to Beyer, globalization has created social, economic or political problems that have given rise to local, particularist responses that tend to be conservative in terms of political outlook. He discusses the Islamic revolution in Iran and new religious Zionism in Israel as examples of particularist, conservative political engagement. These movements champion 'the cultural distinctiveness of one region through a reappropriation of traditional religious antagonistic categories' (1994: 93). They are isolationist, communal, oriented towards local concerns, and are defined by religious goals and values.

One of the explanations that Thomas provides for this apparent global resurgence of religion (of both the ecumenical and the particularist forms) is the 'search for authenticity and development in the developing world' (2005: 41). He also views it as part of a 'revolt against the west' (ibid.: 42) that can be

Plate 9.1 *Christian protestors at 'Occupy', St Paul's Cathedral, London, 27 October 2011: 'what would Jesus do?'*

Source: Joss U (Creative Commons).

divided into three stages: (1) anti-colonial struggles (from the 1940s to the 1960s); (2) the struggle for racial equality (in the 1970s and 1980s); and (3) the struggle for 'cultural liberation', comprising 'the reassertion of traditional and indigenous cultures in the developing world' (ibid.: 42). This later stage became more powerful in the 1990s, since in developing countries the modernizing state in different contexts was perceived to have failed in the provision of development and economic welfare for its citizens. In this context, he argues, the search for authenticity has become more central, where '[a]uthenticity refers to ways of gaining economic prosperity and fashioning political, economic, and social systems that are consistent with a country's moral base, its cultural heritage, and its religious tradition' (ibid.: 11). The search for authenticity, of which the resurgence of religions is a feature, Thomas argues, is 'part of the larger crisis of modernity' (ibid.: 42) and a 'cultural critique of the kind of world that modernity has brought us' (ibid.: 42).

Alongside, and in response to, this resurgence of religion, the past decade has signalled the beginnings of a process of engagement between religions and donor-driven development. However, what does this engagement consist of? Is it a trend that is likely to bring about better development outcomes for people? To what extent is it an agenda that has longevity? Is it perhaps a fad that may not outlive the criticisms and challenges it faces?

Mainstreaming religious and cultural diversity into development?

While many commentators have noted that there has been a 'turn to religion' by mainstream development actors over the past decade, there is a lack of consensus about the extent to which this constitutes an agenda that can contribute towards greater equality and accountability. Some are opposed to this engagement with religion, because of the tendency of religions often to foster divisiveness and inequality. Others strongly hold the view that the mixing of evangelism with aid and development is deeply problematic and that only the most 'secular' religious organizations should be supported. Box 9.1 outlines the work of an Evangelical Christian organization called Samaritan's Purse, which has recently attracted criticism for the way it mixes development with conversion strategies.

As we saw in Chapter 6, many gender activists in particular are concerned that there has been a 'rush to find the religious' in mainstream development policy and practice, stimulated by a 'new global faith agenda' underpinned by western political struggles to deal with rising fundamentalism as well as tendencies to favour supporting FBOs, in both domestic and international settings (Tadros 2010; Patel 2011). The extent to which this critique of engagement with religious actors and organizations should imply an absolute rejection of religious contributions to development is, however, to be questioned. The problem would seem to be linked to the ways in which many development actors engage with religion (e.g. without adequate consideration of the gender consequences), or

---

Box 9.1

Samaritan's Purse

Samaritan's Purse is a controversial Evangelical Christian FBO whose president is Franklin Graham, son of the well-known Christian evangelist Billy Graham. It has attracted particular criticism for its ongoing project Operation Christmas Child. Children and adults in the United Kingdom, Ireland and North America, often through local schools, take part in the project through filling shoeboxes with Christmas treats for disadvantaged children in African, Central Asia and Eastern Europe. This project has received widespread criticism for its practice of including Christian literature within the boxes, typically without the knowledge of those providing the boxes. While the project today claims to, 'where culturally appropriate', 'make available a booklet of Bible stories, which gives a message of hope and an explanation of the true meaning of Christmas – God's gift of His Son, Jesus',[1] this has not calmed the anxiety of critics who believe that combining evangelism and aid is problematic and, moreover, that this literature continues to be handed out in contexts that are non-Christian and alongside other conversion strategies.[2]

---

even the type of religious organization or actors engaged with, rather than to suggest that religion can never be a force for progressive and egalitarian social change. While there are many who are committed to secularism and are therefore likely to view engagement with religions as bound to fail and best avoided, the evidence suggests that this view is not as widespread as it has been in the past and that there is now more openness to thinking about religions and culture as relevant to the development process.

However, this is a long way from any claim that considerations of religion and culture are now 'mainstreamed' into international development processes. In particular, a major critique of the engagement of international development actors with religions in recent years has been that it has tended to be instrumental, using religion where it can help meet pre-defined and western-oriented development goals rather than thinking through how religions themselves contribute towards people's understanding of what counts as development. For instance, the suggestion that mainstream development actors are likely to avoid engaging with organizations that are explicit about their faith motives, are evangelical or appear to have 'extremist' or 'fundamentalist' agendas raises questions about bias and a failure to engage with religions on their own terms. Deneulin and Bano (2009) are highly critical of this instrumentalist approach and instead suggest that because religions are holistic – they encompass the whole lives of the believer – the international development community should deal them with holistically rather than picking and choosing those parts they like and those they do not. Exactly how development actors are to negotiate this terrain, however, is far from clear but, as Deneulin and Bano argue, dialogue could do much to mitigate some of the tensions that currently exist and which arise as a result of rigid views existing across the religious–secular divide (2009: 24–25).

Nonetheless, if Thomas (2005) is correct and we are in the midst of a 'crisis of modernity', within which the West is beginning to question its assumptions about modernization and secularization, then quick, easy and generalizable solutions seem unlikely and instead we should view this questioning as an ongoing process that may take many decades or even generations to unravel. The apparent certainties of western civilization, evolving over many hundreds of years and involving the emergence of the concepts of the religious and the secular, as well as faith in the processes and products of modernity, including secularization and modern neo-liberal markets, and which are at the heart of post-Second World War development project, are unlikely to be dismantled or transformed so easily. Moreover, despite the apparent 'turn to religion' by mainstream development policy and practice, we should not forget that most development research, programmes and projects continue to proceed with no mention of religion and culture.

The extent to which this 'turn to religion' is a trajectory still in full swing is difficult to judge at this stage. It could be the case that it is a temporary fad, and that rather than ironing out the creases and addressing the challenges that the rise of religion poses for development globally, the attention of international development actors will now shift to new and less risky agendas. Such a shift could well push religion out of donor-driven development, as engaging with FBOs and religious leaders loses its appeal; perhaps it will be noted that the types of engagement envisioned and the progress predicted have not been achieved. Whether or not religion remains on international development agendas in the years and decades to come, and in what ways, remains to be seen, but it is clear that religion will continue to shape people's understandings of and approaches to development in many contexts in the Global South. What is also clear is that despite the increased openness to talking about and engaging with religions, this remains a highly complex and contested arena that penetrates to the very core of questions about what development is, what societies should be like, and indeed what it is to be a person. The impossibility of disentangling these questions and implementing solutions to them, from struggles over power, politics and economics, with both secular and religious protagonists, ensures the intractability of an already challenging field.

Rethinking the relationship between religions and international development

In this book we have explored how theories of secularization, which predict the decline of religious belief and practice and/or their relegation to the private sphere, do not seem to fit the evidence provided by the continued or increased significance of religion in many places across the globe, both publicly and privately. Postmodern and postcolonial critiques of secularism (i.e. the view that the distinction between the religious and secular *should be maintained*) argue that secularization theory fails to account for the ways in which religion operates within people's lives, and that to force it out of the public sphere is immoral and impractical, giving rise to distortions in understandings of religious phenomena that have negative impacts for those who embrace them (Asad 1993, 2003). Considering that the avoidance of religion within the mainstream development project since its inception in the post-Second World War period was underpinned by a commitment to secularism and a belief in secularization, there is now a need to rethink the relationship between religions and development from a dominant view that religion hinders development and will eventually disappear to one in which a consideration of religions is given serious attention. This book has also demonstrated that our understandings of what a religion is, as well as predictions about its future, rely upon a particular western perspective. Moreover, the differentiation of the secular from the religious, and hence the invention of the idea of a thing called 'religion', is a relatively recent development emerging from within a European context and is not necessarily replicated everywhere in the same way.

The material covered in this book offers a partial overview of the relationships between religions and international development. While I have focused on several highly relevant themes, some others have received little attention, including religious involvement in health, education and livelihood strategies, as well as considerations of the links between religions and conflict, governance or corruption. Nonetheless, the book has highlighted a series of considerations that are relevant to any attempt within mainstream development studies, policy and practice to engage with religion. I will now summarize these.

One of the most basic requirements that could contribute positively to relationships between religions and international development is religious literacy. This involves not only enhancing one's knowledge of the importance of religions in people's lives in developing contexts but also learning about the beliefs and practices embodied within different religious traditions. Box 9.2 is taken from the blog of Duncan Green, the head of research for Oxfam, in which he discusses the importance of religious literacy for the development sector. He outlines five reasons why he considers this to be important.

However, in the context of moving towards improved processes of dialogue between development and religious actors, an emphasis upon religious literacy for mainstream development actors alone is rather one-sided. Instead, members of different religious traditions working in 'development' should be encouraged to increase their literacy (and hence tolerance) of other religions and denominations, as well as their literacy about the world of development, in order to avoid tarring all international development organizations with the 'World Bank/IMF' brush.

Moving on from the requirement to consider improving different literacies, this book has drawn attention to deeper-level conceptual issues concerning how we understand religions and approach their study. First, religions in developing contexts do not necessarily fit the traditional western Christian-influenced model of what a religion *should be*. For instance, Buddhism does not teach

---

Box 9.2

The need to increase 'religious literacy' in the international development sector

In his blog, Duncan Green has drawn attention to the importance of religious literacy for development policy and practice:

1. Theories of Change: as we think harder about how change happens, religion keeps cropping up, whether it's promoting agency and a sense of 'power within', for example in our work on violence against women, trying to find out how poor people experience poverty, or exploring the internal structures of social movements like the Arab Spring. In all of these, religion plays a crucial part in forming identity and values.

2. Resilience to shocks: whether it's the global financial crisis, or the first few chaotic hours and days after the Haitian earthquake, poor people turn to their churches and mosques for help in an emergency. If we are serious about promoting disaster risk reduction before catastrophe hits, we need to be talking to the institutions that are most relevant to poor people.

3. The growing importance of advocacy: in many developing countries, religion is an important shaper of political space and decisions. Political leaders often defer to religious ones, and getting the faith-based organizations on your side can transform the prospects for change. Just look at the role church-goers played in the successful Jubilee 2000 debt relief campaign, itself a biblical concept.

4. Campaigners are trying to understand better the deep frames that determine how we see the world. Religion plays a vital role in creating those basic frames and values.

5. One of Oxfam's big priorities is gender and 'putting women at the heart of all we do'. Religion plays a huge role in many poor women's lives, and if we want to understand that, we have to leave behind secular sophistries about false consciousness (to caricature, 'religion is really anti-women, it's just that all those women who regularly attend, pray and see their faith as a vital source of wellbeing have . . . been brainwashed [as] to the true extent of their oppression'.)

Source: www.oxfamblogs.org/fp2p/?p=7348 (last accessed 13 September 2012).

---

belief in a divine being, and Hinduism has a greater focus on 'orthopraxy' rather than 'orthodoxy'. Whereas the focus within western religious forms is on the importance of right belief and doctrine, for most Hindus, for instance, belief and doctrine are secondary to practice and how one acts.

Second, focusing on religion as concerned with orthodoxy means that scholars wishing to find out about the nature of different traditions have sometimes tended to prioritize 'official' versions as codified in texts and propounded by religious leaders, rather than the lived religion actually practised by people. If we look for religion only in the obvious places – for example, in the form of religious leaders, institutions or texts – then we are unlikely to get a full picture of patterns of religiosity. It is important not to be overly prescriptive in aligning religious teachings, values and beliefs with what people actually believe in or how they are likely to behave. There is quite often a gulf between 'lived religion' and 'official religion'.

Third, in shifting our focus away from considerations of right belief and doctrinal orthodoxy, we will be more likely to notice that in many contexts the boundaries between religious traditions may actually be more fluid that the 'world religions paradigm' suggests. This 'religious syncretism' or 'hybridity' suggests that a reliance on data that indicate the proportion of a population that practises one religion or another may well be misleading and guilty of reinforcing colonialist categories (Pew Forum 2010).

Fourth, it is often not possible to clearly separate the religious from the secular. People may not think about what they do or what influences them as being 'religious'. This point was addressed, for instance, in Chapter 8, which examined the role of FBOs in development. While some organizations are explicitly 'faith-based' and articulate their identity in this way, others may not acknowledge that religion plays a role in what they do. This could be for political reasons, but also because they exist in a context in which religion influences almost every aspect of life, so it is uncommon to reflect upon where religion is and where it is not. Therefore, in attempting to investigate the influence of religion on development organizations, thinking in terms of 'faith-based' and 'non-faith-based' organizations may not always be the most helpful approach.

As was discussed in Chapter 3, it is now well accepted within contemporary sociology of religion that the shift to viewing religion as focused upon belief, as well as the distinction between different religious traditions, and between the religious and the secular, was facilitated by socio-political changes that occurred in Christian European nations following the upheavals of the Protestant Reformation (Thomas 2005; Asad 2003). This also allowed the emergence of distinct religious traditions (the so-called world religions), which were to be contrasted with the secular as well as with one another (Asad 1993; Beyer 2006). As such, the distinction between the religious and the secular is a rather recent development when placed in the context of the timeline of religious histories. Moreover, while colonialism and globalization have meant that this view of religion is no longer a purely western Christian phenomenon, it is also the case that viewing religion in this way does not always work that well, particularly in non-western contexts where the differentiation between the religious and the secular, as well as between different religions, may be less marked (Beyer 2006). This blurring of the boundaries between the religious and the secular also implies a need to rethink the applicability of the distinction between the private and the public sphere (implicit in theories of secularism and secularization) when attempting to conceptualize the role of religion in modern global societies.

This brings us to the fifth issue, which is concerned with how we might find out about religious attitudes towards development. As we saw in Chapter 4 in particular, religions are highly diverse and contested, both between and within different traditions. There can be as many versions of orthodoxy as there are supporters and promoters of particular positions, and religions can accommodate liberal as well as conservative and fundamentalist interpretations. This has implications for thinking about the themes discussed in this book, such as wealth, poverty and economic development; human rights; gender; and environmentally sustainable development. In bringing religion to the fore in development, the support of religious organizations or the advice of religious leaders is increasingly sought. However, these constituencies often have vested interests in promoting particular worldviews or interpretations of traditions that are not necessarily representative of the needs and interests of others in their communities. In many contexts, religious leaders and organizations are already powerful, and engagement with them by development actors can serve to further strengthen their power and influence. This process can have both positive and negative outcomes, which need to be carefully considered.

Finally, in bringing religion into mainstream development studies, policy and practice it is important not to essentialize poor people as especially religious. Instead, people are influenced by various interconnected and sometimes competing forces, and religion may be one of these. Consideration of religious factors and engagement with religious institutions and actors should be treated as part of a range of approaches to development that are relevant alongside others. A 'religious' approach to development should not be actively pursued as necessarily the best way forward but neither should it be avoided from a commitment to the indivisibility of secularism, modernization and progress.

Summing up

In this concluding chapter the aim has been to address the following question: *What are the implications of the 'rise of religion' for mainstream development studies, policy and practice in the twenty-first century?* We have examined the nature of engagement between religions and development where, to quote from Green, as did Box 9.2, 'engagement can fit anywhere along a spectrum from "find out more" through "dialogue and critical engagement" and "limited issue-based alliances" to "active partnership"'.[3] Compared to previous generations of development studies scholars and development practitioners, for whom to consider religion was sometimes viewed as implying a religious commitment, today we live in an era where the need for an appreciation of the dynamics of religion is taken far more seriously even by those who are not religious. The academic study of religions and international development is still in an early phase, and the shape of the future of engagement between religions and mainstream development actors in practice is far from clear. Nonetheless, while there is no one-size-fits-all model for bringing religion into development (religion can play both a progressive and a regressive role in societies), the continued 'rise of religion' into the twenty-first century is certain and underscores the importance of mainstreaming considerations of religion into development research, policy and practice.

Questions for discussion

1 What are the factors that account for the recent 'turn to religion' by international development actors?

2 What criticisms have been made about the engagement between religions and international development?

3 Which issues would you suggest are important for mainstream development actors to consider in engaging with religion?

Recommended reading

Berger, Peter (ed.) (2000) *The Desecularization of the World: Resurgent Religion and World Politics*, Washington, DC: Ethics and Public Policy Center; Grand Rapids, MI: William B. Eerdmans. This edited volume includes a number of chapters by prominent sociologists of religion. It challenges theories of secularization, suggesting instead that religion may often be strengthened by modernization.

Beyer, Peter (1994) *Religion and Globalization*, London: Sage. In this book, Beyer is interested in the interactions between religions and social change as aspects of processes of globalization. In addition to theoretical reflection, he includes five case studies: the American Christian right; liberation theology movements in Latin America; the Islamic Revolution in Iran; Zionists in Israel; and religious environmentalism.

Deneulin, Séverine and Carole Rakodi (2011) 'Revisiting Religion: Development Studies Thirty Years On', *World Development* 39 (1): 45–54. This paper reassesses the treatment of religion in development studies thirty years after the publication of a special issue of *World Development* entitled 'Religion and Development' (1980).

Special double issue of *Development in Practice* published in July 2012, 22 (5–6) (the table of contents can be viewed at: www.developmentinpractice.org/journals/volume-22-number-5-6) Guest-edited by the director of the DFID-funded Religions and Development programme (RaD: www.religionsanddevelopment.org), Carole Rakodi, with eleven of the sixteen papers based on research undertaken during the programme.

 **Notes**

1 Introduction: religions and development – a new agenda?

1 In this book I will generally use the term 'Global North' rather than 'the West' or 'western', particularly when referring to development in the post-1989 period. The classification that distinguished between 'East' and 'West' became popular during the Cold War. With respect to thinking about different levels of development, western governments referred to the 'West' as the First World, the Eastern Bloc as the Second World and the poorer 'developing' nations as the Third World. Following the collapse of communism in Eastern Europe, the distinction 'Global North' and 'Global South' has prevailed. The term 'Global North' is used to encompass so-called developed countries, including some of the previous Soviet Union states; and 'developing countries' are largely situated in the 'Global South'. Of course this is an imperfect classification since some countries in the Global South are developed (including New Zealand and Australia) and vice versa, but nonetheless this terminology is currently in vogue in development discourse.

2 www.usaid.gov/work-usaid/partnership-opportunities/faith-based-community-organizations/history (last accessed 19 January 2013).

3 http://berkleycenter.georgetown.edu/programs/religion-and-global-development (last accessed 6 September 2012).

4 http://berkleycenter.georgetown.edu/wfdd (last accessed 6 September 2012).

5 http://www.global.ucsb.edu/orfaleacenter/luce/index.html (last accessed 6 September 2012).

6 Ibid.

7 www.religionsanddevelopment.org (last accessed 6 September 2012).

8 www.religion-and-development.nl/home (last accessed 6 September 2012).

2 Approaches to the theory and practice of development: from 'estrangement' to 'engagement' with religions

1 www.jubileedebtcampaign.org.uk/?lid=282 (last accessed 6 September 2012).
2 Ibid.
3 Ibid.
4 http://cases.som.yale.edu/jubilee/index.php?page=1&subMenu (last accessed 6 September 2012).
5 www.jubileedebtcampaign.org.uk/?lid=282 (last accessed 6 September 2012).
6 www.jdcmultifaith.org/abouttheproject.htm (last accessed 6 September 2012).
7 See also http://hdr. undp.org/en/statistics/ (last accessed 19 March 2013).

3 Concepts and theories for studying religions globally

1 http://encarta.msn.com/dictionary_1861710851/secularism.html (last accessed 1 May 2011).

4 Religious approaches to development

1 http://www.islamic-relief.com/InDepth/2-13-poverty.aspx (last accessed 8 September 2012).
2 http://www.islamic-relief.org.uk/Zakat_ul_Fitr.aspx.
3 www.isdb.org:80/irj/portal/anonymous?NavigationTarget=navurl://24de0d5f10 da906da85e96ac356b7af0 (last accessed 8 September 2012).
4 www.isdb.org:80/irj/portal/anonymous?NavigationTarget=navurl://aa8254c9e0a 4110c9f80374f20cb5c6b (last accessed 8 September 2012).
5 www.isdb.org:80/irj/portal/anonymous?NavigationTarget=navurl://46939b3dfc 25365e8588f5ab17627ac5 (last accessed 8 September 2012).
6 Ibid.
7 Email communication, 3 January 2003.

5 Human rights, religions and development

1 www.un.org/en/documents/udhr/history.shtml (last accessed 10 September 2012).
2 www.un.org/documents/ga/res/41/a41r128.htm (last accessed 10 September 2012).
3 United Nations Office of High Commissioner for Human Rights, 'What Is a Rights-Based Approach to Development?', in *Human Rights in Development*. Online, available at: www.unhchr.ch/development/approaches-04.html (last accessed October 2005).
4 www.islamic-relief.com/NewsRoom/6-2-177-ir-takes-part-in-amnesty-international-poverty-debate.aspx (last accessed 10 September 2012).
5 www.vatican.va/holy_father/benedict_xvi/encyclicals/documents/hf_ben-xvi_enc_20090629_caritas-in-veritate_en.html (last accessed 10 September 2012).
6 www1.umn.edu/humanrts/instree/cairodeclaration.html (last accessed 10 September 2012).
7 www.hrcr.org/docs/CEDAW/cedaw5.html (last accessed 10 September 2012).

8 www.law.hku.hk/conlawhk/conlaw/outline/Outline8/Bangkok%20Declaration.htm (last accessed 19 January 2012).

9 See www.oic-oci.org/english/article/human.htm (last accessed 10 September 2012).

10 www.facebook.com/officialSIS/info (last accessed 19 January 2013).

11 http://asiafoundation.org/project/projectsearch.php?country=afghanistan (last accessed 10 September 2012).

6 Gender, religions and development

1 www.unrisd.org/unrisd/website/projects.nsf/(httpProjectsForProgrammeArea-en)/3F3D45E0F8567920C12572B9004180C5?OpenDocument (last accessed 11 September 2012).

7 Environmentalism, religions and development

1 www.earthcharterinaction.org/content/pages/History.html (last accessed 11 September 2012).

2 www.earthcharterinaction.org/content/categories/Religion%20and%20Spirituality/ (last accessed 11 September 2012).

3 www.arcworld.org/about_ARC.asp (last accessed 11 September 2012).

4 www.arcworld.org/about.asp?pageID=2 (last accessed 11 September 2012).

5 www.arcworld.org/arc_and_the_faiths.asp (last accessed 11 September 2012).

6 For instance, see the projects listed at www.arcworld.org/projects.asp?projectID=49 (last accessed 11 September 2012).

7 www.arcworld.org/about.asp?pageID=2 (last accessed 11 September 2012).

8 www.arcworld.org/projects.asp?projectID=50 (last accessed 11 September 2012).

9 http://web.worldbank.org/WBSITE/EXTERNAL/TOPICS/ENVIRONMENT/EXTBIODIVERSITY/0,,contentMDK:20482782~menuPK:1170350~pagePK:148956~piPK:216618~theSitePK:400953,00.html (last accessed 11 September 2012).

10 www.arcworld.org/news.asp?pageID=207 (last accessed 11 September 2012).

11 http://geneva.usmission.gov/2010/04/20/usaid-biodiversity/ (last accessed 11 September 2012).

8 Researching and understanding the role of 'faith-based organizations' (FBOs) in development

1 www.archbishopofcanterbury.org/articles.php/538/archbishops-response-to-DFID-white-paper (last accessed 11 September 2012).

2 www.archbishopofcanterbury.org/articles.php/2539/faith-poverty-and-justice-lambeth-palace-inter-faith-event-with-dfid (last accessed 11 September 2012).

3 http://web.worldbank.org/WBSITE/EXTERNAL/EXTABOUTUS/PARTNERS/EXTDEVDIALOGUE/0,,contentMDK:21955861~menuPK:5555051~pagePK:64192523~piPK:64192458~theSitePK:537298,00.html (last accessed 11 September 2012).

4 http://berkleycenter.georgetown.edu/wfdd (last accessed 11 September 2012).

5 www.unfpa.org/culture/fbo.html (last accessed 11 September 2012).

6 http://transition.usaid.gov/our_work/global_partnerships/fbci/index.html (last accessed 19 December 2011).
7 www.wfd-alliance.org and www.religionsforpeace.org/initiatives/faith-based-forum/ (last accessed 11 September 2012).

9 Conclusion: religions and international development in the twenty-first century

1 www.samaritanspurse.uk.com/operation-christmas-child (last accessed 13 September 2012).
2 www.humanism.org.uk/humanism/humanism-today/humanists-doing/charities/ samaritans-purse (last accessed 13 September 2012).
3 www.oxfamblogs.org/fp2p/?p=7348 (last accessed 13 September 2012).

References

Adams, William (2006) *The Future of Sustainability: Re-thinking Environment and Development in the Twenty-first Century*, Report of the IUCN Renowned Thinkers Meeting, 29–31 January 2006, IUCN, The World Conservation Union. Online, available at: http://cmsdata.iucn.org/downloads/iucn_future_of_sustanability.pdf (last accessed 11 September 2012).

Adamu, Fatima (1999) 'A Double-Edged Sword: Challenging Women's Oppression within Muslim Society in Northern Nigeria', *Gender and Development* 7 (1): 56–61.

Afary, Janet (1997) 'The War against Feminism in the Name of the Almighty: Making Sense of Gender and Muslim Fundamentalism', *New Left Review* (224): 89–110.

Afkhami, Mahnaz (ed.) (1995) *Faith and Freedom: Women's Human Rights in the Muslim World*, London: I. B. Tauris.

Afkhami, Mahnaz and Erika Friedl (eds) (1997) *Muslim Women and the Politics of Participation: Implementing the Beijing Platform*, Syracuse, NY: Syracuse University Press.

Ahmed, Leila (1992) *Women and Gender in Islam: Historical Roots of a Modern Debate*, New Haven, CT: Yale University Press.

Akbarzadeh, Shahramet and Kylie Connor (2008) 'The Organization of the Islamic Conference: Sharing an Illusion', *Middle East Policy* 12 (2): 79–92.

Aldridge, Alan (2007) *Religion in the Contemporary World: A Sociological Introduction*, Cambridge: Polity Press.

Ali, Shaheen Sardar (2000) *Gender and Human Rights in Islam and International Law: Equal before Allah, Unequal before Man?* The Hague: Kluwer Law International.

Alkire, Sabina and Ann Barham (2005) 'Supporting the MDGs: A Faith-Based Movement's Story', *Development* 48 (1): 122–125.

Alley, Kelly (1998) 'Idioms of Degeneracy: Assessing Ganga's Purity and Pollution', in Lance E. Nelson (ed.) *Purifying the Earthly Body of God: Religion and Ecology in Hindu India*, Albany, NY: State University of New York Press, pp. 297–330.

Alley, Kelly D. (2000) 'Separate Domains: Hinduism, Politics and Environmental Pollution', in Christopher Key Chapple and Mary Evelyn Tucker (eds) *Hinduism and Ecology*, Cambridge, MA: Harvard University Press, pp. 355–387.

Alley, Kelly D. (2002) *On the Banks of the Ganga: When Wastewater Meets a Sacred River*, Ann Arbor: University of Michigan Press.

Almond, Philip C. (1988) *The British Discovery of Buddhism*, Cambridge: Cambridge University Press.

Al-Sadr, Muhammad Baqir (2000 [1961]) *Iqtisaduna* (Our economics), London: Bookextra.

Alston, Philip and Mary Robinson (eds) (2005) *Human Rights and Development Towards Mutual Reinforcement*. Oxford: Oxford University Press.

Amin, Sajeda and Sara Hossain (1995) 'Women's Reproductive Rights and the Politics of Fundamentalism: A View from Bangladesh', *American University Law Review* 44 (4): 1319–1343.

Amin, Samir (1976) *Unequal Development: An Essay on the Social Formation of Peripheral Capitalism*, New York: Monthly Review Press.

Amin, Samir (1992) *Empire of Chaos*, New York: Monthly Review Press.

Amirkhanyan, Anna A., Hyun Joon Kim and Kristina T. Lambright (2009) 'Faith-Based Assumptions about Performance: Does Church-Affiliation Matter for Service Quality and Access?', *Nonprofit and Voluntary Sector Quarterly* 38 (3): 490–521.

Anderson, Allan (2004) *An Introduction to Pentecostalism*, Cambridge: Cambridge University Press.

Anderson, Gary M. (1988) 'Mr. Smith and the Preachers: The Economics of Religion in the *Wealth of Nations*', *Journal of Political Economy* 95 (5): 1066–1088.

An-Na'im, Abdullahi Ahmed (ed.) (1992) *Human Rights in Cross-cultural Perspectives: A Quest for Consensus*, Philadelphia: University of Pennsylvania Press.

Apffel-Marglin, Frédérique and P. C. Mishra (1993) 'Sacred Groves: Regenerating the Body, the Land, the Community', in Wolfgang Sachs (ed.) *Global Ecology: A New Arena of Conflict*, Halifax, Nova Scotia: Fernwood Publishing, pp. 197–207.

Apffel-Marglin, Frédérique and Pramod Parjuli (2000) 'Sacred Grove and Ecology: Ritual and Science', in Christopher Key Chapple and Mary Evelyn Tucker (eds) *Hinduism and Ecology: The Intersection of Earth, Sky, and Water*, Cambridge, MA: Harvard University Press, pp. 291–316.

Ariff, Mohamed (1998) 'The Malaysian Economic Experience and Its Relevance for the OIC Member Countries', *Islamic Economic Studies* 6 (1): 1–24.

Ariyaratne, A. T. (1980) *Sarvodaya and Development*, Moratuwa, Sri Lanka: Sarvodaya Press.

Ariyaratne, A. T. (1982) *In Search of Development: The Sarvodaya Shramadana Movement's Effort to Harmonize Tradition with Change*, Moratuwa, Sri Lanka: Sarvodaya Press.

Ariyaratne, A. T. and Joanna Macy (1992) 'The Island of Temple and Tank: Sarvodaya: Self-Help in Sri Lanka', in M. Batchelor and K. Brown (eds) *Buddhism and Ecology*, London: Cassell, pp. 78–86.

Armstrong, Karen (2001) *Islam: A Short History*, Sheffield: Phoenix.

Asad, Talal (1993) *Genealogies of Religion: Discipline and Reasons of Power in Christianity and Islam*, Baltimore: Johns Hopkins University Press.

Asad, Talal (2003) *Formations of the Secular: Christianity, Islam, Modernity*, Stanford, CA: Stanford University Press; London: Eurospan.

Ayub, M. N. (2007) 'Islamic Banking in Pakistan: An Appraisal in Historical Perspective', Religions and Development, Birmingham (unpublished paper).

Baden, Sally and Anne Marie Goetz (1998) 'Who Needs [Sex] When You Can Have [Gender]? Conflicting Discourses on Gender at Beijing ', in Cecile Jackson and Ruth Pearson (eds) *Feminist Visions of Development: Gender Analysis and Policy*, London: Routledge, pp. 19–38.

Balchin, Cassandra (2011) 'Religion and Development: A Practitioner's Perspective on Instrumentalisation', *IDS Bulletin* 42 (1): 15–20.

Balchin, Cassandra (2012) 'Development and Religion: Ambivalent Policy, Grounded Practice', *Open Democracy*, 19 January. Online, available at: www.opendemocracy.net/5050/cassandra-balchin/development-and-religion-ambivalent-policy-grounded-practice (last accessed 10 September 2012).

Banchoff, Thomas and Robert Wuthnow (eds) (2011) *Religion and the Global Politics of Human Rights*, New York: Oxford University Press.

Bandarage, Asoka (1998) *Women, Population and Global Crisis: A Political-Economic Analysis*, London: Zed Books.

Banerjee, Nirmala (2002) 'Between the Devil and the Deep Blue Sea: Shrinking Options for Women in Contemporary India', in Karin Kapadia (ed.) *The Violence of Development: The Politics of Identity, Gender and Social Inequalities in India*, New Delhi: Kali for Women, pp. 43–68.

Banerjee, Nirmala and Jain Devaki (2001) 'Indian Sex Ratios through Time and Space: Development from Women's Perspective', in Vina Mazumdar and N. Krishnaji (eds) *Enduring Conundrum: India's Sex Ratios: Essays in Honour of Asok Mitra*, New Delhi: Centre for Women's Development Studies, pp. 73–119.

Banton, Michael (1966) *Anthropological Approaches to the Study of Religion*, London: Tavistock.

Barnett, Michael and Janice Stein (eds) *Sacred Aid: Faith and Humanitarianism*, Oxford and New York: Oxford University Press.

Barre, Rita Yembilah, Miriam Grant and Dianne Draper (2008) 'The Role of Taboos in Conservation of Sacred Groves in Ghana's Tallensi-Nabdam District, *Social and Cultural Geography* 10 (1): 25–39.

Barro, Robert J. and Rachel McCleary (2001) 'Religion and Economic Growth', NBER Working Paper 9682, Washington, DC: National Bureau of Economic Research.

Barro, Robert J. and Rachel McCleary (2003) 'International Determinants of Religiosity', NBER Working Paper 8931, Washington, DC: National Bureau of Economic Research.

Bartelink, Brenda and Marjo Buitelaar (2006) 'The Challenges of Incorporating Muslim Women's Views into Development Policy: Analysis of a Dutch Action Research Project in Yemen', *Gender and Development* 14 (3): 351–361.

Basham, A. L. (1954) *The Wonder That Was India*, New York: Grove Press.

Basu, Alaka Malwade (1999) 'Fertility Decline and Increasing Gender Imbalance in India, Including a Possible South Indian Turnaround', *Development and Change* 30 (2): 237–263.

Bauer, Joanne R. and Daniel A. Bell (eds) (1999) *The East Asian Challenge for Human Rights*, Cambridge: Cambridge University Press.

Baviskar, Amita (1997) 'Tribal Politics and Discourses of Environmentalism', *Contributions to Indian Sociology* 31 (2): 195–224.

Baviskar, Amita (1999) 'Vanishing Forests, Sacred Trees: A Hindu Perspective on Eco-consciousness', *Asian Geographer* 18 (1–2): 21–31.

Baviskar, Amita (2003) 'Between Violence and Desire: Space, Power, and Identity in the Making of Metropolitan Delhi', *International Social Science Journal* 55 (1): 89–98.

Baviskar, Amita (2004) 'Red in Tooth and Claw? Looking for Class in Struggles over Nature', in Raka Ray and Mary Fainsod Katzenstein (eds) *Social Movements in India: Poverty, Power, Politics*, Lanham, MD: Rowman & Littlefield, pp. 161–178.

Bayes, Jane and Nayereh Tohidi (2001) *Globalization, Gender, and Religion: The Politics of Women's Rights in Catholic and Muslim Contexts*, New York: Palgrave.

Bayly, Susan (1999) *Caste, Society and Politics in India from the Eighteenth Century to the Modern Age*, Cambridge: Cambridge University Press.

Beckford, James (2000) 'Religious Movements and Globalization', in Robin Cohen and Shirin M. Rai (eds) *Global Social Movements*, London, Athlone Press, pp. 165–183.

Beckford, James (2003) *Social Theory and Religion*, Cambridge: Cambridge University Press.

Benthall, Jonathan and Jerome Bellion-Jourdan (2003) *The Charitable Crescent: Politics of Aid in the Muslim World*, London: I. B. Taurus.

Berger, Julia (2003) 'Religious Nongovernmental Organizations: An Exploratory Analysis', *Voluntas: International Journal of Voluntary and Nonprofit Organizations* 14 (1): 15–39.

Berger, Peter L. (1967) *The Sacred Canopy: Elements of a Sociological Theory of Religion*, New York: Anchor Press.

Berger, Peter L. (1974) 'Some Second Thoughts on Substantive versus Functional Definitions of Religion', *Journal for the Scientific Study of Religion* 13 (2): 125–133.

Berger, Peter L. (ed.) (2000) *The Desecularization of the World: Resurgent Religion and World Politics*, Washington, DC: Ethics and Public Policy Center; Grand Rapids, MI: William B. Eerdmans.

Berger, Peter L. and Robert W. Hefner (1998) 'Spiritual Capital in Comparative Perspective'. Online, available at: www.metanexus.net/archive/spiritualcapital researchprogram/pdf/Berger.pdf (last accessed 10 September 2012).

Berman, Marshall (1982) *All That Is Solid Melts into Air: The Experience of Modernity*, New York: Simon & Schuster.

Beyer, Peter (1994) *Religion and Globalization*, London: Sage.

Beyer, Peter (2006) *Religions in Global Society*, London: Routledge.

Bloom, Irene, J. Paul Martin and Wayne L. Proudfoot (eds) (1996) *Religious Diversity and Human Rights*, New York: Columbia University Press.

Boehle, Josef (2010) 'Religious NGOs at the UN and the Millennium Development Goals: An Introduction', *Global Change, Peace and Security* 22 (3): 275–296.

Bopp, Judie (1985) *The Sacred Tree: Reflections on Native American Spirituality*, Twin Lakes, WI: Lotus Light Publications.

Borland, Elizabeth (2004) 'Cultural Opportunity and Tactical Choice in the Argentine and Chilean Reproductive Rights Movements', *Mobilization* 9 (3): 327–339.

Bornstein, Erica (2005) *The Spirit of Development: Protestant NGOs, Morality, and Economics in Zimbabwe*, Stanford, CA: Stanford University Press.

Boserup, Ester (1970) *Women's Role in Economic Development*, London: Allen & Unwin.

Bowen, Donna Lee (1997) 'Abortion, Islam, and the 1994 Cairo Population Conference', *International Journal of Middle East Studies* 29 (2): 161–184.

Bowie, Fiona (1999) *The Anthropology of Religion: An Introduction*, Oxford: Blackwell.

Bradley, Tamsin (2006) *Challenging the NGOs: Religion, Western Discourses and Indian Women*, London: I. B. Tauris.

Bradley, Tamsin (2007) *The Relationships between Religion and Development: Views from Anthropology*, Birmingham: Religions and Development WP 5.

Bradley, Tamsin (2010) *Religion and Gender in the Developing World: Faith-Based Organizations and Feminism in India*, London: I. B. Tauris.

Bradley, Tamsin and Emma Tomalin (2010) 'The Contemporary Dowry Problematic: Exploring the Role of the Study of Religion in Bridging the Gap between Theory and Practice', *Religions of South Asia* 3 (2): 251–274.

Bradley, Tamsin and Emma Tomalin with Mangala Subramaniam (2009) *Dowry: Bridging the Gap between Theory and Practice*, London: Zed Books.

Brandt, Willy (1980) *North–South, a Program for Survival: Report of the Independent Commission on International Development Issues*, London: Pan Books.

Brass, Paul (2006) 'Indian Secularism in Practice', *Indian Journal of Secularism* 9 (1): 115–132.

Brosius, Peter (1999) 'Analyses and Interventions: Anthropological Engagements with Environmentalism', *Current Anthropology* 40 (3): 277–309.

Brown, Usha, Clara Connolly and Pragna Patel (1997) 'Women, "Race" and Culture: Contexts and Campaigns', in Magdalene Ang-Lygate, Chris Corrin and Millsom S. Henry (eds) *Desperately Seeking Sisterhood: Still Challenging and Building*, London: Taylor & Francis, pp. 79–94.

Buchy, Marlène and Bimala Rai (2007) 'Do Women-Only Approaches to Natural Resource Management Help Women? The Case of Community Forestry in Nepal', in Bernadette P. Resurreccion and Rebecca Elmhirst (eds) *Gender and Natural Resource Management: Livelihoods, Mobility and Interventions*, London: Earthscan; Ottawa: International Development Research Centre, pp. 127–150.

Burr, Millard J. and Robert O. Collins (2006) *Alms for Jihad: Charity and Terrorism in the Islamic World*, Cambridge: Cambridge University Press.

Butler, Judith (2008) 'Sexual Politics, Torture, and Secular Time', *British Journal of Sociology* 59 (1): 1–23.

Callicott, J. Baird and Roger T. Aimes (eds) (1991) *Nature in Asian Traditions of Thought: Essays in Environmental Philosophy*, New Delhi: Sri Satguru Publications.

Candland, Christopher (2000) 'Faith as Social Capital: Religion and Community Development in Southern Asia', *Policy Sciences* 33 (3/4): 355–374.

Cardoso, Fernando Henrique and Enzo Faletto (1979) *Dependency and Development in Latin America*, trans. Marjory Mattingly Urquidi, Berkeley: University of California Press.

Carson, Rachel (1962) *Silent Spring*, Boston: Houghton Mifflin.

Casanova, José (1994) *Public Religions in the Modern World*, Chicago: University of Chicago Press.

Casanova, José (2008) 'Public Religions Revisited', in Hent de Vries (ed.) *Religion: Beyond a Concept*, New York: Fordham University Press, pp. 101–119.

Catholic Herald (2011) 'Number of Catholics in the World Grows by 15m in a Year', *Catholic Herald*, 21 February. Online, available at: www.catholicherald.co.uk/news/ 2011/02/21/number-of-catholics-in-the-world-grows-by-15m-in-a-year/ (last accessed 10 September 2012).

Cavanaugh, William T. (1995) 'A Fire Strong Enough to Consume the House: The Wars of Religion and the Rise of the State', *Modern Theology* 11 (4): 397–420.

Centre for Reproductive Rights (2003) Factsheet: 'The Global Gag Rule's Effects on NGOs in 56 Countries'. Online, available at: http://reproductiverights.org/en/ document/the-global-gag-rules-effects-on-ngos-in-56-countries (last accessed 22 December 2011).

Chambers, Robert (1992) 'Rural Appraisal: Rapid, Relaxed and Participatory', IDS Discussion Paper 311, Brighton: Institute of Development Studies, University of Sussex.

Chambers, Robert (1996) *Whose Reality Counts? Putting the First Last*, London: Intermediate Technology Publications.

Chambers, Robert (2004) 'Ideas for Development: Reflecting Forwards', IDS Working Paper 238, Brighton: Institute of Development Studies, University of Sussex. Online, available at: http://opendocs.ids.ac.uk/opendocs/bitstream/handle/123456789/669/ Wp238.pdf (last accessed 13 September 2012).

Chambers, Robert (2005) *Ideas for Development*, London: Earthscan.

Chambre, Susan M. (2001) 'The Changing Nature of "Faith" in Faith-Based Organizations: Secularization and Ecumenism in Four AIDS Organizations in New York City', *Social Service Review* 75 (3): 435–455.

Chan, Joseph (1995) 'The Asian Challenge to Universal Human Rights: A Philosophical Critique', in James Tuck-Hong Tang (ed.) *Human Rights and International Relations in the Asia Pacific Region*, New York: St Martin's Press, pp. 25–38.

Chandran, M. D. Subash and J. Donald Hughes (1997) The Sacred Groves of South India: Ecology, Traditional Communities and Religious Change', *Social Compass* 44 (4): 413–427.

Chapple, Christopher Key (1995) *Nonviolence to Animals, Earth, and Self in Asian Traditions*, New Delhi: Sri Satguru Publications.

Chapple, Christopher Key (1998) 'Toward an Indigenous Indian Environmentalism', in Lance E. Nelson (ed.) *Purifying the Earthly Body of God: Religion and Ecology in Hindu India*, Albany, NY: State University of New York Press, pp. 13–37.

Chapple, Christopher Key and Mary Evelyn Tucker (eds) (2000) *Hinduism and Ecology: The Intersection of Earth, Sky, and Water*, Cambridge, MA: Harvard Divinity School, Center for the Study of World Religions/Harvard University Press.

Chaves, Mark (2002) 'Religious Organizations: Data Resources and Research Opportunities', *American Behavioral Scientist* 45 (10): 1523–1549.

Chenery, H., M. S. Ahluwalia, C. L. G. Bell, J. H. Duloy and R. Jolly (1974) *Redistribution with Growth*, Oxford: Oxford University Press.

Chidester, David (1996) *Savage Systems: Colonialism and Comparative Religion in Southern Africa*, Charlottesville: University Press of Virginia.

Clarke, Gerard (2006) 'Faith Matters: Faith-Based Organizations, Civil Society and International Development', *Journal of International Development* 18 (6): 835–848.

Clarke, Gerard (2007) 'Agents of Transformation? Donors, Faith-Based Organisations and International Development', *Third World Quarterly* 28 (1): 77–96.

Clarke, Gerard (2008) 'Faith-Based Organizations and International Development: An Overview', in Gerard Clarke and Michael Jennings (eds) *Development, Civil Society and Faith-Based Organizations: Bridging the Sacred and the Secular*, Basingstoke, UK: Palgrave Macmillan, pp. 17–45.

Clarke, Gerard and Michael Jennings (2008a) Introduction, in Gerard Clarke and Michael Jennings (eds) *Development, Civil Society and Faith-Based Organizations: Bridging the Sacred and the Secular*, Basingstoke, UK: Palgrave Macmillan, pp. 1–16.

Clarke, Gerard and Michael Jennings (eds) (2008b) *Development, Civil Society and Faith-Based Organizations: Bridging the Sacred and the Secular*, Basingstoke, UK: Palgrave Macmillan.

Clarke, Matthew (2011) *Development and Religion: Theology and Practice*, Cheltenham, UK and Northampton, MA: Edward Elgar Publishing.

Clifford, James and George E. Marcus (eds) (1986) *Writing Culture: The Poetics and Politics of Ethnography*, Berkeley: University of California Press.

Cochrane, J. R., B. Schmid and T. Cutts (eds) (2011) *When Religion and Health Align: Mobilizing Religious Health Assets for Transformation*, Pietermaritzburg, South Africa: Cluster Publications.

Cohen, Susan A. (2001) 'Global Gag Rule: Exporting Antiabortion Ideology at the Expense of American Values', *The Guttmacher Report on Public Policy* 4 (3). Online, available at: www.guttmacher.org/pubs/tgr/04/3/gr040301.html (last accessed 22 December 2011).

Commoner, Barry (1971) *The Closing Circle: Nature, Man and Technology*, New York: Knopf.

Connolly, Peter (1999) *Approaches to the Study of Religion*, London: Cassell.

Connolly, William E. (2006) 'Europe: A Minor Tradition', in David Scott and Charles Hirschkind (eds) *Powers of the Secular Modern: Talal Asad and His Interlocutors*, Stanford, CA: Stanford University Press, pp. 75–92.

Conway, Dennis and Nikolas Heynen (2008) 'Dependency Theories: From ECLA to Andre Gunder Frank and Beyond', in Vandana Desai and Robert B. Potter (eds) *The Companion to Development Studies*, 2nd edn, London: Hodder Education, pp. 92–96.

Cooey, Paula M., William R. Eakin and Jay B. McDaniel (eds) (1991) *After Patriarchy: Feminist Transformations of the World Religions*, Maryknoll, NY: Orbis.

Cornwall, Andrea and Sarah White (eds) (2000*) Men, Masculinities and Development: Politics, Policies and Practice*, Brighton: Institute of Development Studies, University of Sussex.

Cornwall, Andrea, Jerker Edström and Alan Greig (eds) (2011) *Men and Development: Politicizing Masculinities*, London: Zed Books.

Couture, D. (2003) 'Women's Rights and Religion: An Analysis of Some Islamic and Catholic Discourses', *Studies in Religion/Sciences Religieuses* 32 (1–2): 5–18.

Cox, Harvey (1996) *Fire from Heaven: The Rise of Pentecostal Spirituality and the Reshaping of Religion in the Twenty-first Century*, London: Cassell.

Crane, Barbara B. and Jennifer Dusenberry (2004) 'Power and Politics in International Funding for Reproductive Health: The US Global Gag Rule', *Reproductive Health Matters* 12 (24): 128–137.

Cumaraswamy, Dato' Param (1997) 'Human Rights and Universality: The Universal Declaration of Human Rights Is It Universal?' Online, available at: www.hrsolidarity.net/mainfile.php/1997vol07no07/2082/ (last accessed 12 September 2012).

Daly, Mary (1973) *Beyond God the Father: Toward a Philosophy of Women's Liberation*, Boston: Beacon Press.

Dasgupta, Subhaya (1977) *Hindu Ethos and the Challenge of Change*, New Delhi: Arnold-Heinemann.

DasGupta Sherma, Rita (1998) 'Sacred Immanence: Reflections of Ecofeminism in Hindu Tantra', in Lance E. Nelson (ed.) *Purifying the Earthly Body of God: Religion and Ecology in India*, Albany, NY: State University of New York Press, pp. 89–132.

Davies, Brian (2004) *An Introduction to the Philosophy of Religion*, 3rd edn, Oxford: Oxford University Press.

Deb, Debal (2007) 'Sacred Groves of West Bengal: A Model of Community Forest Management', Understanding Livelihood Impacts of Participatory Forest Management Implementation in India and Nepal, Working Paper 8, Overseas Development Group, University of East Anglia, UK. Online, available at: www.uea.ac.uk/dev/People/staffresearch/ospringate-baginskiresearch/PFM-Nepal-India/8-sacred-groves-west-bengal (last accessed 11 September 2012).

de Beaufort, Fleur, Ingemund Hägg and Patrick van Schie (eds) (2008) *Separation of Church and State in Europe with Views on Sweden, Norway, the Netherlands, Belgium, France, Spain, Italy, Slovenia and Greece*, European Liberal Forum. Online, available at: www.liberalforum.eu/tl_files/publications/Separation%20of%20church%20and%20state%20in%20Europe.pdf (last accessed 11 September 2012).

De Cordier, Bruno (2009) 'Faith-Based Aid, Globalisation and the Humanitarian Frontline: An Analysis of Western-Based Muslim Aid Organisations', *Disasters* 33 (4): 608–628.

Deneulin, Séverine with Masooda Bano (2009) *Religion in Development: Rewriting the Secular Script*, London: Zed Books.

Deneulin, Séverine and Carole Rakodi (2011) 'Revisiting Religion: Development Studies Thirty Years On', *World Development* 39 (1): 45–54.

Desai, Vandana (2008) 'The Role of Non-governmental Organizations', in Vandana Desai and Robert B. Potter (eds) *The Companion to Development Studies*, 2nd edn, London: Hodder Education, pp. 525–530.

Desai, Vandana and Robert B. Potter (eds) (2008) *The Companion to Development Studies*, 2nd edn, London: Hodder Education.

DFID (2008) *The Gender Manual: A Practical Guide*, London: Department for International Development. Online, available at: http://new.uneca.org/Portals/ngm/Documents/Gender%20Mainstreaming/dfid-gender-manual-2008.pdf (last accessed 22 December 2011).

DFID (2009) *Eliminating World Poverty: Building our Common Future*, White Paper, London: Department for International Development. Online, available at: www.official-documents.gov.uk/document/cm76/7656/7656.pdf (last accessed 11 September 2012).

DFID (2012) *Faith Partnership Principles: Working Effectively with Faith Groups to Fight Global Poverty*, London: Department for International Development. Online, available at: www.archbishopofcanterbury.org/canterbury//data/files/resources/2539/Faith-Partnership-Principles.pdf (last accessed 11 September 2012).

Dikirr, Patrick M. (2008) 'The Challenges of Indigenizing Africa's Environmental Conservation Goals', *Journal of Pan African Studies* 2 (3): 81–99.

Dillon, Michele E. (ed.) (2003) *Handbook of the Sociology of Religion*, Cambridge: Cambridge University Press.

Dillon, Michele E. (2009) 'The Sociology of Religion', in Bryan Turner (ed.) *The New Blackwell Companion to Social Theory*, 3rd edn, Oxford: Blackwell, pp. 409–427.

Dillon, Michele E. (2010) *Introduction to Sociological Theory: Theorists, Concepts, and Their Applicability in the Twenty-first Century*, Malden, MA: Wiley-Blackwell.

Dinham, Adam, Richard Furbey and Vivien Lowndes (2009) *Faith in the Public Realm*, Bristol: Policy Press.

Dirks, Nicholas B. (2001) *Castes of Mind: Colonialism and the Making of Modern India*, Princeton, NJ: Princeton University Press.

Donnelly, E. A. (2002) 'Proclaiming Jubilee: The Debt and Structural Adjustment Network', in Sanjeev Khagram, James V. Riker and Kathryn Sikkink (eds) *Restructuring World Politics: Transnational Social Movements, Networks and Norms*, Minneapolis: University of Minnesota Press, pp. 155–180.

Donno, Daniela and Bruce M. Russett (2004) 'Islam, Authoritarianism, and Female Empowerment: What Are the Linkages?', *World Politics* 56 (4): 582–607.

Dreze, Jean and Amartya Sen (2002) *India: Development and Participation*, Oxford: Oxford University Press.

Drumbl, Mark A. (2004) 'Rights, Culture, and Crime: The Role of Rule of Law for the Women of Afghanistan', *Columbia Journal of Transnational Law* 42 (2): 349–390.

Dumont, Louis (1970) *Homo Hierarchicus: The Caste System and Its Implications*, trans. [from the French] Mark Sainsbury, London: Weidenfeld & Nicolson.

Durkheim, Emile (1995 [1912]) *Elementary Forms of Religious Life*, New York: Free Press.

Dutta, K. and A. Robinson (1997) *Selected Letters of Rabindranath Tagore*, Cambridge: Cambridge University Press.

Eade, Deborah (2002) Preface, in Deborah Eade (ed.) *Development and Culture*, Oxford: Oxfam GB in Association with World Faiths Development Dialogue, pp. ix–xiv.

Ehrlich, Paul (1968) *The Population Bomb*, New York: Ballantine Books.

El Dawla, Aida Seif (2000) 'Reproductive Rights of Egyptian Women: Issues for Debate', *Reproductive Health Matters* 8 (16): 45–54.

Engel, J. Ronald (2005) 'Biosphere Reserves and World Heritage Sites', in Bron Taylor (ed.) *Encyclopedia of Religion and Nature*, London: Continuum, pp. 192–194.

Engineer, Asghar Ali (2004) *The Rights of Women in Islam*, Elgin, IL: New Dawn Press.

Escobar, Arturo (1995) *Encountering Development: The Making and Unmaking of the Third World*, Princeton, NJ: Princeton University Press.

Feng, Wang (1996) 'A Decade of One Child Policy: Achievements and Implications', in Alice Goldstein and Wang Feng (eds) *China: The Many Facets of Demographic Change*, Boulder, CO: Westview Press, pp. 97–120.

Fenn, R. K. (2003) *The Blackwell Companion to Sociology of Religion*, Oxford: Blackwell.

Fernandes, Ruben Cesar (1992) 'A Night for the Earth', *The New Road: The Magazine of WWF's Conservation and Religion Network*, June–September: 7.

Fitzgerald, Timothy (1990) 'Hinduism and the "World Religion" Fallacy', *Religion* 20 (2): 101–118.

Fitzgerald, Timothy (2000) *The Ideology of Religious Studies*, New York: Oxford University Press.

Flood, Gavin D. (1996). *An Introduction to Hinduism*, Cambridge: Cambridge University Press.

Flood, Gavin D. (1999) *Beyond Phenomenology: Rethinking the Study of Religion*, London: Cassell.

Foley, Rebecca (2004) 'Muslim Women's Challenges to Islamic Law: The Case of Malaysia', *International Feminist Journal of Politics* 6 (1): 53–84.

Foltz, Richard, Azizan Baharuddin and Frederick M. Denny (eds) (2004) *Islam and Ecology: A Bestowed Trust*, Cambridge, MA: Harvard Divinity School, Center for the Study of World Religions/Harvard University Press.

Foreign Policy (2007) 'The List: The World's Fastest-Growing Religions', *Foreign Policy*, 14 May. Online, available at: www.foreignpolicy.com/articles/2007/05/13/the_list_the_worlds_fastest_growing_religions (last accessed 12 January 2012).

Fox, Judith (2005) 'Secularization', in John R. Hinnells (ed.) *The Routledge Companion to the Study of Religion*, London: Routledge, pp. 291–305.

Frank, Andre Gunder (1967) *Capitalism and Underdevelopment in Latin America: Historical Studies of Chile and Brazil*, New York: Monthly Review Press.

Freeman, John Richardson (1994) 'Forests and the Folk: Perceptions of Nature in the Swidden Regimes of Highland Malabar', Pondy Papers in Social Sciences 15, Pondicherry: French Institute.

Furseth, Inger and Pal Repstad (2006) *An Introduction to the Sociology of Religion: Classical and Contemporary Perspectives*, Aldershot, UK: Ashgate.

Fuss, Diana (1989) *Essentially Speaking: Feminism, Nature and Difference*, New York: Routledge.

Gadgil, Madhav and Subash M. D. Chandran (1992) 'Sacred Groves', *India International Centre Quarterly* 19 (1–2): 183–187.

Galeano, Eduardo (2000) *Upside Down: A Primer for the Looking Glass World*, New York: Metropolitan Books.

Geaves, Ron (1998) 'The Borders between Religions: A Challenge to the World Religions Approach to Religious Education'. *British Journal of Religious Education* 21 (1): 20–31.

Geertz, Clifford (1973) *The Interpretation of Cultures: Selected Essays*, New York: Basic Books.

Gerami, Shahin and Melodye Lehnerer (2001) 'Women's Agency and Household Diplomacy: Negotiating Fundamentalism', *Gender and Society* 15 (4): 556–573.

Ghanea, Nazila (ed.) (2009) *Religion and Human Rights* (4 vols), London: Routledge.

Gibelman, Margaret and Sheldon R. Gelman (2002) 'Should We Have Faith in Faith-Based Social Services? Rhetoric versus Realistic Expectations', *Nonprofit Management and Leadership* 13 (1): 49–65.

Giddens, Anthony (1991) *The Consequences of Modernity*, Stanford, CA: Stanford University Press.

Gilman, Nils (2007) *Mandarins of the Future: Modernization Theory in Cold War America*, Baltimore: Johns Hopkins University Press.

Githitho, Anthony N. (2003) 'The Sacred Mijikenda Kaya Forests of Coastal Kenya and Biodiversity Conservation', in Cathy Lee and Thomas Schaaf (eds) *The Importance of Sacred Natural Sites for Biodiversity Conservation: Proceedings of the International Workshop held in Kunming and Xishuangbanna Biosphere Reserve, People's Republic of China, 17–20 February 2003*, Paris: United Nations Educational, Scientific and Cultural Organization (UNESCO), pp. 27–35. Online, available at: http://unesdoc.unesco.org/images/0013/001333/133358e.pdf (last accessed 12 September 2012).

Glazier, Stephen D. (1998) 'Anthropology of Religion', in William H. Swatos Jr (ed.) *Encyclopedia of Religion and Society*. Online, available at: http://hirr.hartsem.edu/ency/anthropology.htm (last accessed 11 September 2012).

Gold, Daniel (1991) 'Organized Hinduisms: From Vedic Truth to Hindu Nation', in Martin Marty and Scott R. Appleby (eds) *Fundamentalisms Observed*, Chicago: University of Chicago Press, pp. 531–593.

Gorski, Philip S. (2003) 'Historicizing the Secularization Debate: An Agenda for Research', in Michele Dillon (ed.) *Handbook of the Sociology of Religion*, Cambridge: Cambridge University Press, pp. 110–122.

Gottlieb, Roger S. (1996) *This Sacred Earth: Religion, Nature, Environment*, New York: Routledge.

Green, Cathy, Susan Joekes and Melissa Leach (1998) 'Questionable Links: Approaches to Gender in Environmental Research and Policy', in Cecile Jackson and Ruth Pearson (eds) *Feminist Visions of Development*, London: Routledge, pp. 259–283.

Green, Maia, Claire Mercer and Simeon Mesaki (2013) 'Faith in Forms: Civil Society Evangelism and Development in Tanzania', *Development in Practice* 22 (5–6): 721–734.

Greenhalgh, Susan (2001) 'Fresh Winds in Beijing: Chinese Feminists Speak Out on the One-Child Policy and Women's Lives', *SIGNS* 26 (3): 847–886.

Grillo, R. D. and R. L. Stirrat (eds) (1997) *Discourses of Development: Anthropological Perspectives*, Oxford: Berg.

Grills, Nathan (2009) 'The Paradox of Multilateral Organizations Engaging with Faith-Based Organizations', *Global Governance* 15 (4): 505–520.

Grimes, Seamus (1998) 'From Population Control to "Reproductive Rights": Ideological Influences in Population Policy', *Third World Quarterly* 19 (3): 375–393.

Guha, Ramachandra (1989) 'Radical American Environmentalism and Wilderness Preservation: A Third World Critique', *Environmental Ethics* 11 (1): 71–83.

Guha, Ramachandra and Juan Martínez-Alier (1997) *Varieties of Environmentalism: Essays North and South*, London: Earthscan.

Gutiérrez, Gustavo (1973 [1971]) *A Theology of Liberation: History, Politics, Salvation*, Maryknoll, NY: Orbis Books.

Hackett, Rosalind I. J. (1995) 'The Gospel of Prosperity in West Africa', in Richard H. Roberts (ed.) *Religion and the Transformations of Capitalism: Comparative Approaches*, London: Routledge, pp. 199–214.

Hajjar, Lisa (2004) 'Religion, State Power, and Domestic Violence in Muslim Societies: A Framework for Comparative Analysis', *Law and Social Inquiry* 29 (1): 1–38.

Hamm, Brigitte I. (2001) 'A Human Rights Approach to Development', *Human Rights Quarterly* 23 (4): 1005–1031.

Haq, Mahbub ul (1976) *The Third World and the International Economic Order*, Washington, DC: Overseas Development Council.

Haq, Mahbub ul (1995) *Reflections on Human Development*, Oxford: Oxford University Press.

Harper, Ian R. and Samuel Gregg (eds) (2010) *Christian Theology and Market Economics*, Cheltenham, UK: Edward Elgar.

Harris, Ian (1991) 'How Environmentalist Is Buddhism?', *Religion* 21 (2): 101–114.

Hasna, Fadia (2003) 'Islam, Social Traditions and Family Planning', *Social Policy and Administration* 37 (2): 181–197.

Hawley, John S. and Wayne L. Proudfoot (1994) Introduction, in John S. Hawley (ed.) *Fundamentalism and Gender*, New York: Oxford University Press, pp. 3–44.

Haynes, Jeffrey (2007) *Religion and Development: Conflict or Cooperation?* Basingstoke, UK: Palgrave Macmillan.

Haynes, Jeffrey (2008) *Development Studies*, Cambridge: Polity Press.

Haynes, Jeffrey (2011) 'Religion, Secularization and Politics: A Postmodern Conspectus', in Jeffrey Haynes, *Religion, Politics and International Relations: Selected Essays*, London: Routledge, pp. 12–30.

Hefferan, Tara, Julie Adkins and Laurie Occhipinti (2009) 'Faith-Based Organisations, Neo-liberalism and Development: An Introduction', in Tara Hefferan, Julie Adkins and Laurie Occhipinti (eds) *Bridging the Gaps: Faith-Based Organizations, Neoliberalism, and Development in Latin America and the Caribbean*, Lanham, MD: Lexington Books, pp. 1–34.

Helie, Anissa (2004) 'Holy Hatred', *Reproductive Health Matters* 12 (23): 120–124.

Hennelly, Alfred T. (1990) *Liberation Theology: A Documentary History*, Maryknoll, NY: Orbis Books.

Hessel, Dieter T. and Rosemary Radford Ruether (eds) (2000) *Christianity and Ecology: Seeking the Well-Being of Earth and Humans*, Cambridge, MA: Harvard Divinity School, Center for the Study of World Religions/Harvard University Press.

Hill, Michael (2000) '"Asian Values" as Reverse Orientalism: Singapore', *Asia Pacific Viewpoint* 41 (2): 177–190.

Hoksbergen, Ronald G. (2005) 'Building Civil Society through Partnership: Lessons from a Case Study of the Christian Reformed World Relief Committee', *Development in Practice* 15 (1): 16–27.

Hoksbergen, Ronald G. and Lowell Ewert (eds) (2002) *Local Ownership: Global Change: Will Civil Society Save the World?* Monrovia, CA: World Vision Publications.

Holenstein, Anne-Marie (2005) *Role and Significance of Religion and Spirituality in Development Co-operation*, Bern: Swiss Agency for Development and Cooperation SDC.

Holt, Mack P. (1993) 'Putting Religion Back into the Wars of Religion', *French Historical Studies* 18 (2): 524–551.

Hopkins, Adrienne and Kirit Patel (2006) 'Reflecting on Gender Equality in Muslim Contexts in Oxfam GB', *Gender and Development* 14 (3): 423–435.

Howland, Courtney W. (ed.) (2001) *Religious Fundamentalisms and the Human Rights of Women*, New York: Palgrave.

Hulme, David (2007) *The Making of the Millennium Development Goals: Human Development Meets Results-Based Management in an Imperfect World*, Brooks World Poverty Institute Working Paper 16. Online, available at: www.bwpi.manchester.ac. uk/resources/Working-Papers/bwpi-wp-1607.pdf (last accessed 12 September 2012).

Huntington, Samuel P. (1993) 'The Clash of Civilizations?', *Foreign Affairs* 72 (3): 22–49.

Iannaccone, Lawrence R. (1998) 'An Introduction to the Economics of Religion', *Journal of Economic Literature* 36 (3): 1465–1496.

Iannaccone, Lawrence R. and Jonathan Klick (2003) 'Spiritual Capital: An Introduction and Literature Review', Paper presented to Spiritual Capital Planning Meeting, Cambridge, MA, 9–10 October. Online, available at: http://www.metanexus.net/ archive/spiritualcapitalresearchprogram/pdf/review.pdf (last accessed 3 January 2012).

Inden, Ronald (1986) 'Orientalist Constructions of India', *Modern Asian Studies* 20 (3): 401–446.

Islamic Relief (2008) *Definitions of Poverty*. Online, available at: http://www.islamic-relief.com/indepth/downloads/Islamic%20Relief%20-%20Definitions%20of%20 Poverty%20-%20Jan08.pdf (last accessed 13 March 2013).

Jackson, Cecile (1994) 'Gender Analysis and Environmentalisms', in Michael R. Redclift and Ted Benton (eds) *Social Theory and the Global Environment*, London: Routledge, pp. 113–149.

Jackson, Cecile (2001) 'Gender, Nature and Trouble with Anti-dualism', in Alaine Low and Soraya Tremayne (eds) *Women as Sacred Custodians of the Earth? Women, Spirituality and the Environment*, New York: Berghahn Books, pp. 23–44.

Jackson, Cecile and Ruth Pearson (eds) (1998) *Feminist Visions of Development, Gender Analysis and Policy*, London: Routledge.

Jackson, Paul and Christiane Fleischer (2007) *Religion and Economics: A Literature Review*, Birmingham: Religions and Development WP 3.

Jacobsen, Michael and Ole Bruun (eds) (2000) *Human Rights and Asian Values: Contesting National Identities and Cultural Representations in Asia*, Richmond, Surrey, UK: Curzon.

Jaffrelot, Christophe (2008) 'Hindu Nationalism and the Social Welfare Strategy', in Gerard Clarke and Michael Jennings (eds) *Development, Civil Society and Faith-Based Organizations: Bridging the Sacred and the Secular*, Basingstoke, UK: Palgrave Macmillan, pp. 240–259.

Jahan, Ronauq (1995) *The Elusive Agenda: Mainstreaming Women in Development*, London: Zed Books.

Jamal, Amina (2005) 'Feminist "Selves" and Feminism's "Others": Feminist Representations of Jamaat-e-Islami Women in Pakistan', *Feminist Review* 81: 52–73.

James, Rick (2009) 'What Is Distinctive about FBOs? How European FBOs Define and Operationalise Their Faith', Oxford: INTRAC Praxis Paper 22. Online, available at: http://dspace.cigilibrary.org/jspui/bitstream/123456789/24075/1/What%20is%20dist inctive%20about%20FBOs.pdf?1 (last accessed 3 January 2012).

James, Rick (2011) 'Handle with Care: Engaging with Faith-Based Organisations in Development', *Development in Practice* 21 (1): 109–117.

Jayaraj, Dhairiyarayar and Sreenivasan Subramanian (2004) 'Women's Wellbeing and the Sex Ratio at Birth: Some Suggestive Evidence from India', *Journal of Development Studies* 40 (5): 91–119.

Jeavons, Thomas H. (2004) 'Religious and Faith-Based Organizations: Do We Know One When We See One?', *Nonprofit and Voluntary Sector Quarterly* 33 (1): 140–145.

Jodkha, Surinder and Pradyumna Bora (2009) *Mapping Faith-Based Development Activities in Contemporary Maharashtra, India*, Birmingham: Religions and Development WP 28.

Johnson, Byron R., Ralph Brett Tompkins and Derek Webb (2002) *Objective Hope: Assessing the Effectiveness of Faith-Based Organizations: A Review of the Literature*, Center for Research on Religion and Urban Civil Society, University of Pennsylvania. Online, available at: www.manhattan-institute.org/pdf/crrucs_objective_hope.pdf (last accessed 3 January 2012).

Jung, Kim Dae (1994) 'Is Culture Destiny? The Myth of Asia's Anti-Democratic Values', *Foreign Affairs* 73 (6): 189–194.

Junhong, Chu (2001) 'Prenatal Sex Determination and Sex Selective Abortion in Rural Central China', *Population and Development Review* 27 (2): 259–281.

Kabeer, Naila and Ramya Subrahmanian (eds) (2000) *Institutions, Relations and Outcomes: A Framework and Case Studies for Gender Aware Planning*, London: Zed Books.

Kalam, M. A. (1996) 'Sacred Groves in Kodagu District of Karnataka (South India): A Socio-historical Study', Pondicherry: Pondy Papers in Social Sciences 21.

Kalu, Ogbu U. (2003) 'Safiyya and Adamah: Punishing Adultery with Sharia Stones in Twenty-first-century Nigeria, *African Affairs* 102 (408): 389–408.

Kane, Gillian (2008) 'Abortion Law Reform in Latin America: Lessons for Advocacy', *Gender and Development* 16 (2): 361–375.

Kardam, Nüket (2005) *Turkey's Engagement with Global Women's Human Rights Norms*, Aldershot, UK: Ashgate.

Katsouris, Christina (1999) 'More Debt-distressed African Countries Could Qualify for HIPC Relief: Creditors Making "Major Changes" in Debt Relief for Poor Countries. Africa Recovery'. *United Nations Department of Public Information*, 13 (2–3): 11–15. Online, available at: www.un.org/en/africarenewal/subjindx/ subpdfs/132debt. pdf (last accessed 6 September 2012).

Kausikan, Bilahari (1995–1996) 'An East Asian Approach to Human Rights', *Buffalo Journal of International Law* 2: 263–283.

Kausikan, Bilahari (1997) 'Governance That Works', *Journal of Democracy* 8 (2): 24–33.

Kazimov, Necva B. (2003) 'Egypt's Reservations in the United Nations Convention on the Elimination of Discrimination against Women and Women's Rights in Egypt', *American Journal of Islamic Social Sciences* 20 (3–4): 1–45.

Keddie, Nikki (1994) 'The Rights of Women in Contemporary Islam', in Leroy S. Rouner (ed.) *Human Rights and the World's Religions*, Notre Dame, IN: University of Notre Dame Press, pp. 76–93.

Keown, Damien, Charles S. Prebish and Wayne Rollen Husted (eds) (1998) *Buddhism and Human Rights*, Richmond, Surrey, UK: Curzon Press.

Khare, R. S. (1998) 'The Issue of "Right to Food" among the Hindus: Notes and Comments', *Contributions to Indian Sociology* 32 (2): 253–278.

Kiely, Ray (1999) 'The Last Refuge of the Noble Savage: A Critical Assessment of Post-Developmental Theory', *The European Journal of Development Research* 11(1): 30–55.

Killingley, Dermot (1993) *Rammohun Roy in Hindu and Christian Tradition*, Newcastle upon Tyne: Grevatt & Grevatt.

Kim, Kirsteen (2007) *Concepts of Development in the Christian Traditions*, Birmingham: Religions and Development WP 16.

King, Anna S. (1996) 'Spirituality: Transformation and Metamorphosis', *Religion* 26 (4): 343–351.

King, Richard (1999) *Orientalism and Religion: Postcolonial Theory, India and 'The Mystic East'*, London: Routledge.

Kirmani, Nida (2012) 'The Role of Religious Values and Beliefs in Charitable and Development Organisations in Karachi and Sindh, Pakistan', *Development in Practice* 22 (5–6): 735–748.

Kirmani, Nida and Sarah Zaidi (2010) *The Role of Faith in the Charity and Development Sector in Karachi and Sindh*, Birmingham: Religions and Development WP 50.

Kishwar, Madhu (no date) 'From Manusmriti to Madhusmriti: Flagellating a Mythical Enemy'. Online, available at: www.infinityfoundation.com/ECITmythicalframeset. htm (last accessed 3 January 2012).

Klasen, Stephan (1994) 'Missing Women Reconsidered', *World Development* 22 (7): 1061–1071.

Klasen, Stephan and Claudia Wink (2003) 'Missing Women: Revisiting the Debate', *Feminist Economics* 9 (2–3): 263–299.

Klostermaier, Klaus (1994) *A Survey of Hinduism*, Albany, NY: State University of New York Press.

Knott, Kim (2000) *Hinduism: A Very Short Introduction*, Oxford: Oxford Paperbacks.

Kothari, Uma (ed.) (2005) *A Radical History of Development Studies: Individuals, Institutions and Ideologies*, Cape Town: David Philip; London: Zed Books.

Kroessin, Mohammed Ralf (2007) 'Islamic Charities and the "War on Terror": Dispelling the Myths', *Islamic Relief Humanitarian Exchange Magazine* 38. Online, available at: www.odihpn.org/report.asp?id=2890 (last accessed 12 September 2012).

Kroessin, Mohammed Ralf (2008) *Concepts of Development in 'Islam': A Review of Contemporary Literature and Practice*, Birmingham: Religions and Development WP 20.

Kroessin, Mohammed Ralf (2011) 'A Genealogy of the Islamic Development Discourse: Underlying Assumptions and Policy Implications from a Development Studies Perspective', Paper given at the Eighth International Conference on Islamic Economics and Finance, Doha, Qatar. Online, available at http://conference.qfis.edu. qa/app/media/293 (last accessed 19 January 2013).

Kung, Hans and Karl-Josef Kuschel (1993) *A Global Ethic: The Declaration of the Parliament of the World's Religions*, London: SCM Press.

Lachenmann, Gudrun (2004) 'Female Spaces in West African Societies', *Peripherie* 95 (24): 322–340.

Laws, John (2003) 'Beyond Rights', *Oxford Journal of Legal Studies* 23 (2): 265–280.

Laws of Manu (1886), trans. George Bühler (Sacred Books of the East, vol. 25). Online, available at: www.sacred-texts.com/hin/manu.htm (last accessed 11 September 2012).

Leach, Melissa (2007) 'Earth Mother Myths and Other Ecofeminist Fables: How a Strategic Notion Rose and Fell', *Development and Change* 38 (1): 67–85.

Lee, Cathy and Thomas Schaaf (2003) 'Annex 1', in Cathy Lee and Thomas Schaaf (eds) *The Importance of Sacred Natural Sites for Biodiversity Conservation: Proceedings of the International Workshop held in Kunming and Xishuangbanna Biosphere Reserve, People's Republic of China, 17–20 February 2003*, Paris: United Nations

Educational, Scientific and Cultural Organization (UNESCO), pp. 163–165. Online, available at: http://unesdoc.unesco.org/images/0013/001333/133358e.pdf (last accessed 12 September 2012).

Lee, Martin (1998) *Human Rights – Perspectives: Call for a New Asian Rights Charter*. Online, available at: www.hrsolidarity.net/mainfile.php/1998vol08no12/1870/?print =yes (last accessed 12 September 2012).

Leurs, Robert, Peter Tumaini-Mungu and Abu Mvungi (2011) *Mapping the Development Activities of Faith-Based Organizations in Tanzania*, Birmingham: Religions and Development WP 58.

Lewis, David and Kanji, Nazneen (2009) *Non-governmental Organizations and Development*, London: Routledge.

Li, Jiali (1995) 'China's Family Planning Program: How Well Has It Worked? A Case Study of Hebei Province 1979–1988', *Population and Development Review* 21 (3): 563–585.

Lincoln, Bruce (2003) *Holy Terrors: Thinking about Religion after September 11,* Chicago: University of Chicago Press.

Lindsey, Tim (2004) 'Indonesia: Devaluing Asian Values, Rewriting Rule of Law', in Randall Peerenboom (ed.) *Asian Discourses of Rule of Law: Theories and Implementation in Twelve Asian Countries, France and the U.S.*, London: RoutledgeCurzon, pp. 286–323.

Lipsky, Alyson B. (2006) 'Evaluating the Strength of Faith: Potential Comparative Advantages of Faith-Based Organizations Providing Health Services in Sub-Saharan Africa', *Public Administration and Development* 31 (1): 25–36.

Little, David (1999) 'Rethinking Human Rights: A Review Essay on Religion, Relativism and Other Matters', *Journal of Religious Ethics* 27 (1): 151–177.

McCutcheon, Russell T. (1997) *Manufacturing Religion: The Discourse on Sui Generis Religion and the Politics of Nostalgia*, New York: Oxford University Press.

McCutcheon, Russell T. (1999) *The Insider/Outsider Problem in the Study of Religion: A Reader*, London: Cassell.

McCutcheon, Russell T. (2001) *Critics not Caretakers: Redescribing the Public Study of Religion*, Albany, NY: State University of New York Press.

Macdonald, Mandy (ed.) (1994) *Gender Planning in Development Agencies: Meeting the Challenge*, Oxford: Oxfam.

McGrath, Alister E. (2006) *Christianity: An Introduction*, Malden, MA: Blackwell.

McGrath, Alister E. (2011) *Theology: The Basics*, Oxford: Blackwell.

MacGregor, Sherilyn (2007) *Beyond Mothering Earth: Ecological Citizenship and the Politics of Care*, Vancouver: University of British Columbia Press.

McKean, Lise (1996) *Divine Enterprise: Gurus and the Hindu Nationalist Movement*, Chicago: University of Chicago Press.

Mandani, Mahmoos (2004) *Good Muslim, Bad Muslim: America, the Cold War, and the Roots of Terror*, New York: Pantheon.

Marks, Stephen P. (2001) 'The Human Rights Framework for Development: Five Approaches' (revised and expanded version of a talk delivered at the UNDP-sponsored Second Global Forum on World Development, held in Rio de Janeiro on 9–10 October 2000).

Marks, Stephen P. (2004) 'The Human Right to Development: Between Rhetoric and Reality', *Harvard Human Rights Journal* 17: 137–168.

Maroney, Eric (2006) *Religious Syncretism*, St Albans, UK: SCM Press.

Marshall, Katherine and Lucy Keough (eds) (2004) *Mind, Heart, and Soul in the Fight against Poverty*, Washington, DC: World Bank.

Marshall, Katherine and Richard Marsh (eds) (2003) *Millennium Challenges for Development and Faith Institutions*, Washington, DC: World Bank Publications,

Marshall, Mandy and Nigel Taylor (2006) 'Tackling HIV and AIDS with Faith-Based Communities: Learning from Attitudes on Gender Relations and Sexual Rights within Local Evangelical Churches in Burkina Faso, Zimbabwe, and South Africa', *Gender and Development* 14 (3): 363–374.

Martin, David (1979) *A General Theory of Secularization*, New York: Harper & Row.

Martin, David (2002) *Pentecostalism*, Oxford: Blackwell.

Martin, David (2005) *On Secularization: Towards a Revised General Theory*, Aldershot, UK: Ashgate.

Marty, Martin E. and Appleby, R. Scott (1991) *Fundamentalisms Observed*, Chicago: University of Chicago Press.

Maududi, Sayyid Abul A'la (1970) *Ma'ashiyat-e Islam* (Economic system of Islam), Lahore: Islamic Publications.

Mawdsley, Emma (2004) 'India's Middle Classes and the Environment', *Development and Change* 35 (1): 79–103.

Mayer Ann Elizabeth (1994) 'Universal versus Islamic Human Rights: A Clash of Cultures or a Clash with a Construct?', *Michigan Journal of International Law* 15: 307–404.

Meadows, Donella H., Dennis L. Meadows, Jørgen Randers, and William W. Behrens III (1972) *The Limits to Growth*, New York: Universe Books.

Melkote, Srinivas R. and H. Leslie Steeves (2001) *Communication for Development in the Third World: Theory and Practice for Empowerment*, 2nd edn, New Delhi: Sage Publications India.

Mercer, Claire (2002) 'NGOs, Civil Society and Democratization in the Developing World: A Critical Review of the Literature', *Progress in Development Studies* 2 (1): 5–22.

Merchant, Carolyn (1992) *Radical Ecology: The Search for a Livable World*, New York: Routledge.

Mernissi, Fatima (1975) *Beyond the Veil: Male–Female Dynamics in a Modern Muslim Society*, New York: Schenkman.

Miller, David (1999) 'Modernity in Hindu Monasticism: Swami Vivekananda and the Ramakrishna Movement', *Journal of Asian Studies* 34 (1): 111–126.

Milton, Kay (1996) *Environmentalism and Cultural Theory: The Role of Anthropology in Environmental Discourse*, London: Routledge.

Minor, R. N. (1998) *The Religious, the Spiritual, and the Secular: Auroville and Secular India*, Albany: State University of New York Press.

Moaddel, Mansoor (1998) 'Religion and Women: Islamic Modernism versus Fundamentalism', *Journal for the Scientific Study of Religion* 37 (1): 108–130.

Moaddel, Mansoor (2002) 'The Study of Islamic Culture and Politics: An Overview and Assessment', *Annual Review of Sociology* 28: 359–386.

Moddie, A. D. (1968) *The Brahmanical Culture and Modernity*, Bombay: Asia Publishing House.

Moghadam, Valerie (2000) 'Transnational Feminist Networks: Collective Action in an Era of Globalization', *International Sociology* 15 (1): 57–85.

Molyneux, Maxine (1985) 'Mobilization without Emancipation? Women's Interests, the State and Revolution in Nicaragua', *Feminist Studies* 2 (2): 227–254.

Momsen, Janet (2010) *Gender and Development*, 2nd edn, London: Routledge.

Moser, Caroline (1993) *Gender Planning and Development: Theory, Practice and Training*, London: Routledge.

Munson, Henry (2005) 'Fundamentalism', in John R. Hinnells (ed.) *The Routledge Companion to the Study of Religion*, London: Routledge, pp. 337–354.

Næss, Arne (1973) 'The Shallow and the Deep, Long Range Ecology Movement: A Summary', *Inquiry* 16: 95–100.

Narayan, Deepa with Raj Patel, Kai Schafft, Anne Rademacher and Sarah Koch-Schulte (2000a) *Voices of the Poor: Can Anyone Hear Us?* New York: Oxford University Press for the World Bank.

Narayan, Deepa, Robert Chambers, Meera Kaul Shah and Patti Petesch (2000b) *Voices of the Poor: Crying Out for Change*, New York: Oxford University Press for the World Bank.

Narayan, Shriman (1970) *Relevance of Gandhian Economics*, Ahmedabad, India: Navajivan Publishing House.

Nelson, Lance E. (1998) 'The Dualism of Nondualism: Advaita Vedānta and the Irrelevance of Nature', in Lance E. Nelson (ed.) *Purifying the Earthly Body of God: Religion and Ecology in India*, Albany, NY: State University of New York Press, pp. 61–88.

Nelson, Robert H. (2003) 'Environmental Colonialism "Saving" Africa from Africans', *Independent Review* 8 (1): 65–86.

Niblock, Timothy and Rodney Wilson (eds) (1999) *The Political Economy of the Middle East*, vol. 3: *Islamic Economics*, Cheltenham, UK: Edward Elgar.

Nishimuko, Mikako (2008) 'The Role of Faith-Based Organizations in Building Democratic Process: Achieving Universal Primary Education in Sierra Leone', *International Journal of Social Sciences* 3 (3): 172–179.

Nkurunziza, Emmanuel (2007) *An Overview of Development Studies: Background Paper*, Birmingham: Religions and Development WP 2.

Nussbaum, Martha C. (2000) *Women and Human Development: The Capabilities Approach*, Cambridge: Cambridge University Press.

Obermeyer, Carla Makhlouf (1994) 'Reproductive Choice in Islam: Gender and State in Iran and Tunisia', *Studies in Family Planning* 25 (1): 41–51.

Oberoi, H. (1994) *The Construction of Religious Boundaries: Culture, Identity and Diversity in the Sikh Tradition*, Delhi: Oxford University Press.

Occhipinti, Laurie A. (2005) *Acting on Faith: Religious Development Organizations in Northwestern Argentina*, Lanham, MD: Lexington Books.

OHCHR (1996) *Fact Sheet No.2 (Rev.1), The International Bill of Human Rights*. Online, available at: http://www.ohchr.org/Documents/Publications/FactSheet2Rev.1en.pdf (last accessed 12 March 2013).

OHCHR (2006) *Frequently Asked Questions on a Human Rights-based Approach to Development Cooperation*. Online, available at: http://www.ohchr.org/Documents/Publications/FAQen.pdf (last accessed 12 March 2013).

Olivier, Jill and Quentin Wodon (eds) (2011) *Strengthening Faith-Inspired Health Engagement in Africa*, Washington, DC: World Bank.

Olson, Elizabeth (2008) 'Common Belief, Contested Meanings: Development and Faith-Based Organizational Culture', *Royal Dutch Geographical Society* 99 (4): 393–405.

O'Neill, Onora (2002) Reith Lectures 2002: Lecture 2 – 'Trust and Terror'. Online, available at: www.bbc.co.uk/radio4/reith2002/lecture2.shtml (last accessed 12 September 2012).

Oommen, T. K. (2004) *Nation, Civil Society and Social Movements: Essays in Political Sociology*, New Delhi: Sage.

Oslington, Paul (2008) 'Christianity's Post-Enlightenment Contribution to Economic

Thought', in Ian Harper and Samuel Gregg (eds) *Christian Morality and Market Economics: Theological and Philosophical Perspectives*, Aldershot, UK: Edward Elgar, pp. 60–76. Also available at http://apps.acu.edu.au/staffdirectory/file_folder/7400c43850d97a910f9e598b3b5ac92f.pdf (last accessed 4 January 2012).

Owens, Michael Leo (2007) *God and Government in the Ghetto: The Politics of Church–State Collaboration in Black America*, Chicago: University of Chicago Press.

Padgett, Tim (2012) 'Sorry, Rome, U.S. Catholics Are More Like Melinda Gates', 12 July. Online, available at: http://ideas.time.com/2012/07/12/sorry-rome-us-catholics-more-like-melinda-gates/ (last accessed 11 September 2012).

Palmer, Martin with Victoria Finlay (2003) *Faith in Conservation: New Approaches to Religions and the Environment*, Washington, DC: World Bank. Online, available at: http://siteresources.worldbank.org/INTBIODIVERSITY/214584-1112712965549/20480342/FaithInConservationNewApproachesPreface2003.pdf (last accessed 3 January 2012).

Pandaya, Meenal (1998) 'Ethical Economics', *Hinduism Today*, March. Online, available at: www.hinduismtoday.com/modules/smartsection/item.php?itemid=4729 (last accessed 12 September 2012).

Pani, Narendar (2002) *Inclusive Economics: Gandhian Method and Contemporary Policy*, New Delhi: Sage.

Parpat, Jane L. (2002) 'Lessons from the Field: Rethinking Empowerment, Gender and Development from a Post- (Post-?) Development Perspective', in Kriemhild Saunders (ed.) *Feminist Post-development Thought: Rethinking Modernity, Post-colonialism and Representation*, London: Zed Books, pp. 41–56.

Parsons, Diane and William B. Jonte-Pace (eds) (2001) *Religion and Psychology: Mapping the Terrain*. London: Routledge.

Patel, Pragna (2011) 'Women Migrants and Faith Organisations: Changing Regimes of Gender, Religion and Race in London', *Feminist Review* 97 (1): 142–150.

Patel, Sujay and Amin Muhammad Gadit (2008) 'Karo-Kari: A Form of Honour Killing in Pakistan', *Transcultural Psychiatry* 45 (4): 683–694.

Pearson, Ruth (2000) 'Rethinking Gender Matters in Development', in Tim Allen and Alan Thomas (eds) *Poverty and Development into the 21st Century*, Milton Keynes: Open University in association with Oxford University Press, pp: 383–402.

Pearson, Ruth and Emma Tomalin (2008) 'Intelligent Design? A Gender Sensitive Interrogation of Religion and Development', in Gerard Clarke and Michael Jennings (eds) *Development, Civil Society and Faith-Based Organizations: Bridging the Sacred and the Secular*, Basingstoke, UK: Palgrave, pp. 46–71.

Pedersen, Poul (1995) 'Nature, Religion and Cultural Identity: The Religious Environmentalist Paradigm', in Ole Bruun and Arne Kalland (eds) *Asian Perceptions of Nature: A Critical Approach*, Richmond, Surrey, UK: Curzon Press, pp. 258–276.

Peerenboom, Randall Paul (2000) 'Human Rights and Asian Values: The Limits of Universalism', *China Review International* 7 (2): 295–320.

Peerenboom, Randall Paul (2003) 'Beyond Universalism and Relativism: The Evolving Debates about "Values in Asia"', *Indiana International and Comparative Law Review* 14: 1–86.

PEPFAR (2005) *The President's Emergency Plan for AIDS Relief: Community and Faith-Based Organizations*. Online, available at: http://www.state.gov/documents/organization/54420.pdf (last accessed 11 March 2013).

Perry, Michael J. (1997) 'Are Human Rights Universal? The Relativist Challenge and Related Matters', *Human Rights Quarterly* 19 (3): 461–509.

Pew Forum (2006) *Spirit and Power: A 10-Country Survey of Pentecostals*. Online, available at: http://pewforum.org/Christian/Evangelical-Protestant-Churches/Spirit-and-Power.aspx (last accessed 20 March 2013).

Pew Forum (2008) *U.S. Religious Landscape Survey. Religious Beliefs and Practices: Diverse and Politically Relevant*. Online, available at: http://religions.pewforum.org/pdf/report2-religious-landscape-study-full.pdf (last accessed 12 September 2012).

Pew Forum (2010) *Tolerance and Tension: Islam and Christianity in Sub-Saharan Africa, Executive Summary*. Online, available at: http://pewforum.org/executive-summary-islam-and-christianity-in-sub-saharan-africa.aspx (last accessed 20 March 2013).

Pew Research Center (2002) 'U.S. Stands Alone in Its Embrace of Religion among Wealthy Nations, Overview', 19 December. Online, available at: www.pewglobal.org/2002/12/19/among-wealthy-nations/ (last accessed 12 September 2012).

Piron, Laure-Hélène (2002) *The Right to Development: A Review of the Current State of the Debate for the Department for International Development*. Commissioned by DFID's Social Development Department. Online, available at: www.odi.org.uk/resources/download/1562.pdf (last accessed 12 September 2012).

Piron, Laure-Hélène with Tammie O'Neil (2005) *Integrating Human Rights into Development: A Synthesis of Donor Approaches and Experiences*. Prepared for the OECD DAC Network on Governance (GOVNET). Online, available at: http://odi.org.uk/sites/odi.org.uk/files/odi-assets/publications-opinion-files/4404.pdf (last accessed 12 September 2012).

Plumwood, Val (1986) 'Ecofeminism: An Overview and Discussion of Positions and Arguments', *Australian Journal of Philosophy* 64: 120–138.

Poole, Eve (2010) *The Church on Capitalism: Theology and the Market*, Basingstoke, UK: Palgrave Macmillan.

Putnam, Robert (1995) 'Bowling Alone: America's Declining Social Capital', *Journal of Democracy* 6 (1): 65–78.

Rakodi, Carole (2011) *A Guide to Analyzing the Relationships between Religion and Development*, Birmingham: Religions and Development WP 67.

Ramadan, Tariq (2010) 'Good Muslim, Bad Muslim'. Online, available at: www.newstatesman.com/religion/2010/02/muslim-religious-moderation.

Rathgeber, Eva M. (1990) 'WID, WAD, GAD: Trends in Research and Practice', *Journal of Developing Areas* 24 (4): 489–502.

Rauf, Abdul (2007) 'Pan-Islamism and the North West Frontier Province of British India (1897–1918)', *Perceptions*, Winter: 21–42.

Rawi, Mariam (2004) 'Betrayal', *Reproductive Health Matters* 12 (23): 116–119.

Rawls, John (1996) *Political Liberalism*, New York: Columbia University Press.

Razavi, Shahra (2009) Foreword, in José Casanova and Anne Philips, 'A Debate on the Public Role of Religion and its Social and Gender Implications', Gender and Development Programme Paper 5, UNRISD, pp. iii–v.

Razavi, Shahra and Anne Jenichen (2010) 'The Unhappy Marriage of Religion and Politics: Problems and Pitfalls for Gender Equality', *Third World Quarterly* 31 (6): 833–850.

Reilly, Niamh (2011) 'Rethinking the Interplay of Feminism and Secularism in a Neo-secular Age', *Feminist Review* 97 (1): 5–31.

Religions for Peace (2005) 'Faith in Action: Working toward the Millennium Development Goals (an Action Toolkit for Religious Leaders and Communities)'. Online, available at: www.worldvolunteerweb.org/fileadmin/docdb/mdg_faith_toolkit.pdf (last accessed 21 December 2011).

Renteln, Alison Dundes (1990) *International Human Rights: Universalism versus Relativism*, Newbury Park, CA: Sage.

Rist, Gilbert (2002) *The History of Development: From Western Origins to a Global Faith*, London: Zed Books.

Rodrigues, Hillary (2006) *Introducing Hinduism*, New York: Routledge.

Rose, Susan (2005) 'Going Too Far? Sex, Sin and Social Policy', *Social Forces* 84 (2): 1207–1232.

Rouner, Leroy S. (ed.) (1994) *Human Rights and the World's Religions*, Notre Dame, IN: University of Notre Dame Press.

Rowlands, Jo (1998) 'A Word of the Times, but What Does It Mean? Empowerment in the Discourse and Practice of Development', in Haleh Afshar (ed.) *Women and Empowerment: Illustrations of the Third World*, Basingstoke, UK: Macmillan, pp. 11–34.

Rozario, Santi (2009) 'Dowry in Rural Bangladesh: An Intractable Problem?', in Tamsin Bradley and Emma Tomalin with Mangala Subramaniam (eds) *Dowry: Bridging the Gap between Theory and Practice*, London: Zed Books, pp. 29–58.

Rudnyckyj, Daromir (2011) *Spiritual Economies: Islam, Globalization, and the Afterlife of Development*, Ithaca, NY: Cornell University Press.

Ruether, Rosemary Radford (1983) *Sexism and God-Talk: Towards a Feminist Theology*, London: SCM Press.

Runzo, Joseph, Nancy M. Martin and Arvind Sharma (eds) (2003) *Human Rights and Responsibilities in the World Religions*, Oxford: Oneworld Publications.

Ruthven, Malise (2000) *Islam: A Very Short Introduction*, Oxford: Oxford Paperbacks.

Rylaarsdam, Coert (1968) 'Poverty and the Poor in the Bible', *Pastoral Psychology* 19 (3): 13–24.

Sachs, Wolfgang (1992) *The Development Dictionary: A Guide to Knowledge as Power*, London: Zed Books.

Said, Edward (1979) *Orientalism*, New York: Vintage Books.

Said, Edward (1993) *Culture and Imperialism*. New York: Knopf.

Said, Edward (1997) *Covering Islam: How the Media and the Experts Determine How We See the Rest of the World*, New York: Vintage.

Said, Edward (2001) 'The Clash of Ignorance', *The Nation*, 22 October. Online, available at: www.thenation.com/article/clash-ignorance (last accessed 12 September 2012).

Sánchez Fuentes, María Luisa, Jennifer Paine and Brooke Elliott-Buettner (2008) 'The Decriminalisation of Abortion in Mexico City: How Did Abortion Rights Become a Political Priority?', *Gender and Development* 16 (2): 345–360.

Schaaf, Thomas (2003) 'UNESCO's Experience with the Protection of Sacred Natural Sites for Biodiversity Conservation', in Cathy Lee and Thomas Schaaf (eds) *The Importance of Sacred Natural Sites for Biodiversity Conservation: Proceedings of the International Workshop held in Kunming and Xishuangbanna Biosphere Reserve, People's Republic of China, 17–20 February 2003*, Paris: United Nations Educational, Scientific and Cultural Organization (UNESCO), pp. 13–20. Online, available at: http://unesdoc.unesco.org/images/0013/001333/133358e.pdf (last accessed 3 January 2012).

Schussler-Fiorenza, Elizabeth (1983) *In Memory of Her: A Feminist Theological Reconstruction of Christian Origins*, London: SCM Press.

Sechzer, Jeri Altneu (2004) 'Islam and Woman: Where Tradition Meets Modernity: History and Interpretations of Islamic Women's Status', *Sex Roles* 51 (5–6): 263–272.

Seed, Jan (1999) 'A History of Gender Training in Oxfam', in Fenella Porter, Ines Smyth and Caroline Sweetman (eds) *Gender Works: Oxfam Experience in Policy and Practice*, Oxford: Oxfam, pp. 311–317.

Seguino, Stephanie (2011) 'Help or Hindrance? Religion's Impact on Gender Inequality', *World Development* 39 (8): 1308–1321.

Selinger, Leah (2004) 'The Forgotten Factor: The Uneasy Relationship between Religion and Development', *Social Compass* 51 (4): 523–543.

Sen, Amartya (1985) *Commodities and Capabilities*, Cambridge: Cambridge University Press.

Sen, Amartya (1990) 'More than 100 Million Women Are Missing', *New York Review of Books* 37 (20), 20 December. Online, available at: http://ucatlas.ucsc.edu/gender/sen100m.html (last accessed 12 September 2012).

Sen, Amartya (1992) 'Missing Women', *British Medical Journal* 304 (6827): 587–588. Online, available at: www.bmj.com/content/304/6827/587.full.pdf (last accessed 13 September 2012).

Sen, Amartya (1993) 'The Economics of Life and Death', *Scientific American* 268: 40–47. Online, available at: www.wfu.edu/~heckeljc/ecn150/LifeandDeath.pdf (last accessed 13 September 2012).

Sen, Amartya (1996) 'Thinking about Human Rights and Asian Values', *Human Rights Dialogue* 1 (4). Online, available at: www.carnegiecouncil.org/publications/archive/dialogue/1_04/articles/519.html (last accessed 13 September 2012).

Sen, Amartya (1997) 'Human Rights and Asian Values', Sixteenth Morgenthau Memorial Lecture on Ethics and Foreign Policy, New York: Carnegie Council on Ethics and International Affairs. Online, available at: www.mtholyoke.edu/acad/intrel/sen.htm (last accessed 13 September 2012).

Sen, Amartya (1999) *Development as Freedom*, Oxford: Oxford University Press.

Sen, Amartya (2003) 'Missing Women – Revisited', *British Medical Journal* 327 (7427): 1297–1298.

Sharma, Arvind (1999) Letter, *Journal of Religious Ethics* 27 (3): 539–549.

Sharma, Arvind (2004) *Hinduism and Human Rights: A Conceptual Approach*, Oxford: Oxford University Press.

Sharma, Arvind (2010) *Hindu Narratives on Human Rights*, Santa Barbara, CA: Greenwood.

Sharma, R. (1997) *Gandhian Economics*, New Delhi: Deep & Deep Publications.

Shaw, Rosalind (1995) 'Feminist Anthropology and the Gendering of Religious Studies', in Ursula King (ed.) *Religion and Gender*, Oxford: Blackwell, pp. 65–76.

Shiva, Vandana (1988) *Staying Alive: Women, Ecology and Survival in India*, New Delhi: Kali for Women.

Sholkamy, Hania (2011) 'Creating Conservatism or Emancipating Subjects? On the Narrative of Islamic Observance in Egypt', *IDS Bulletin* 42 (1): 47–55.

Sidaway, James D. (2008) 'Post-development', in Vandana Desai and Robert B. Potter (eds) *The Companion to Development Studies*, 2nd edn, London: Hodder Education, pp. 16–19.

Sider, Ronald J. and Heidi R. Unruh (2004) 'Typology of Religious Characteristics of Social Service and Educational Organizations and Programs', *Nonprofit and Voluntary Sector Quarterly* 33 (1): 109–134.

Simon, David (2008) 'Neoliberalism, Structural Adjustment and Poverty Reduction strategies', in Vandana Desai and Robert B. Potter (eds) *The Companion to Development Studies*, 2nd edn, London: Hodder Education, pp. 86–92.

Singer, Milton (1956) 'Cultural Values in India's Economic Development', *Annals of the American Academy of Political and Social Science* 305: 81–91.

Singh, Gurharpal, Heather Marquette and Namawu Alhassan Alolo (2007) *Political*

Science, Religion and Development: A Literature Review, Birmingham: Religions and Development WP 7.

Singh, Nikky-Guninder Kaur (1993) *The Feminine Principle in the Sikh Vision of the Transcendent*, Cambridge: Cambridge University Press.

Sinha, Subir, Shubhra Gururani and Brian Greenberg (1997) 'The "New Traditionalist" Discourse of Indian Environmentalism', *Journal of Peasant Studies* 24 (3): 65–99.

Slaby, Michael C. and AtKisson, Alan (2007) *The Earth Charter and Religion: Toward an ECI Program on Religion and Sustainability*, Earth Charter International, Center for Strategy and Communications, Stockholm. Online, available at: www.earthcharterinaction.org/pdfs/ecireligionstrat2007.pdf (last accessed 12 September 2012).

Sleboda, Jennifer (2001) 'Islam and Women's Rights Advocacy in Malaysia', *Asian Journal of Women's Studies* 7 (2): 94–136.

Smart, Ninian (1973) *The Science of Religion and the Sociology of Knowledge: Some Methodological Questions*, Princeton, NJ: Princeton University Press.

Smart, Ninian (1978) *The Phenomenon of Religion*, London: Mowbrays.

Smart, Ninian (1996) *Dimensions of the Sacred: An Anatomy of the World's Beliefs*, London: HarperCollins.

Smith, Jonathan Z. (1982) *Imagining Religion: From Babylon to Jonestown*, Chicago: University of Chicago Press.

Smith, Stephen R. and Michael R. Sosin (2001) 'The Varieties of Faith-Related Agencies', *Public Administration Review* 61 (6): 651–670.

Smith, Wilfred Cantwell (1991 [1962]) *The Meaning and End of Religion*, Minneapolis, MN: Fortress Press.

Smyth, Ines (1998) 'Gender Analysis of Family Planning: Beyond the "Feminist vs. Population Control" Debate', in Cecile Jackson and Ruth Pearson (eds) *Feminist Visions of Development: Gender Analysis and Policy*, London: Routledge, pp. 217–238.

Soares, Benjamin F (2006) 'Islam in Mali in the Neoliberal Era', *African Affairs* 105 (418): 77–95.

Sookhdeo, Rosemary (2008) *Secrets behind the Burqa*, McLean, VA: Isaac Publishing.

Spivak, G. C. (1987) *In Other Worlds: Essays in Cultural Politics*, New York: Methuen.

Srinivas, M. N. (1962) *Caste in Modern India and Other Essays*, London: Asia Publishing House.

Stabile, Carole A. and Deepa Kumar (2005) 'Unveiling Imperialism: Media, Gender and the War on Afghanistan', *Media, Culture and Society* 27 (5): 765–782.

Stambach, Amy (2005) 'Rallying the Armies or Bridging the Gulf: Questioning the Significance of Faith-Based Educational Initiatives in a Global Age', *Indiana Journal of Global Legal Studies* 12 (1): 205–226.

Stanislaus, L. (1999) *The Liberative Mission of the Church among Dalit Christians*, Delhi: ISPCK.

Streeten, Paul, Shahid Javed Burki, Mahbub ul Haq, Norman Hicks and Francis Stewart (1981) *First Things First: Meeting Basic Human Needs in Developing Countries*, Oxford: Oxford University Press.

Sturgeon, Noel (1997) *Ecofeminist Natures: Race, Gender, Feminist Theory and Political Action*, New York: Routledge.

Sumner, Andy and Michael Tribe (2008) *International Development Studies: Theories and Methods in Research and Practice*, London: Sage.

Sunder Rajan, Rajeswari (2003) *The Scandal of the State: Women, Law, and Citizenship in Postcolonial India*, Durham, NC: Duke University Press.

Sung, Jung Mo (2007) *Desire, Market, Religion (Reclaiming Liberation Theology)*, London: SCM Press.

Sutcliffe, Steven J. (2004) *Religion: Empirical Studies: A Collection to Mark the 50th Anniversary of the British Association for the Study of Religion*, Aldershot, UK: Ashgate.

Sylvester, G. (1999) 'Development Studies and Postcolonial Studies: Disparate Tales of the "Third World"', *Third World Quarterly* 20 (4): 703–721.

Tadros, Mariz (2010) 'Faith-Based Organizations and Service Delivery: Some Gender Conundrums', UNRISD, Gender and Development Programme Paper 11. Online, available at: www.unrisd.org/unrisd/website/document.nsf/8b18431d756b7085 80256b6400399775/592137c50475f6a8c12577bd004fb5a0/$FILE/Tadros.pdf (last accessed 13 September 2012).

Tadros, Mariz (2011) 'Introduction: Gender, Rights and Religion at the Crossroads', *IDS Bulletin* 42 (1): 1–9.

Tang, James Tuck-Hong (ed.) (1995) *Human Rights and International Relations in the Asia Pacific Region*, New York: St Martin's Press.

Taringa, Nisbert (2006) 'How Environmental Is African Traditional Religion?', *Exchange* 35 (2): 191–214.

Taylor, Bron, Iain S. Maclean and Heather Eaton (2005) 'United Nations "Earth Summits"', in Bron Taylor (ed.) *Encyclopedia of Religion and Nature*, London: Continuum, pp. 1680–1683.

Taylor, Charles (1999) 'Conditions of an Unforced Consensus on Human Rights', in Joanne R. Bauer and Daniel A. Bell (eds) *The East Asian Challenge for Human Rights*, Cambridge: Cambridge University Press, pp. 124–146.

Taylor, Charles (2007) *A Secular Age*, Cambridge, MA: Harvard University Press.

Taylor, Michael (1995) *Not Angels but Agencies: The Ecumenical Response to Poverty – a Primer*, London: SCM Press.

Taylor, Michael (2003) *Christianity, Poverty and Wealth*, London: SPCK.

ter Haar, Gerrie (ed.) (2011) *Religion and Development: Ways of Transforming the World*, London: C. Hurst.

Thomas, Alan (2000) 'Poverty and the "End of Development"', in Tim Allen and Alan Thomas (eds) *Poverty and Development into the 21st Century*, Milton Keynes: Open University in association with Oxford University Press, pp. 3–22.

Thomas, Owen C. and Ellen K. Wondra (2002) *Introduction to Theology*, 3rd edn, Harrisburg, PA: Morehouse Publishing.

Thomas, Scott M. (2005) *The Global Resurgence of Religion and the Transformation of International Relations: The Struggle for the Soul of the Twenty-first Century*, New York: Palgrave Macmillan.

Tilley, John J. (2000) 'Cultural Relativism', *Human Rights Quarterly* 22 (2): 501–547.

Timberg, Thomas A. (2000) *Islamic Banking in Indonesia*, Jakarta: Partnership for Economic Growth.

Tomalin, Emma (2004a) 'Bio-divinity and Biodiversity: Perspectives on Religion and Environmental Conservation in India', *Numen* 51 (3): 265–295.

Tomalin, Emma (2004b) 'Deep Ecology', in Tim Forsyth (ed.) *Encyclopedia of International Development*, London: Routledge.

Tomalin, Emma (2006a) 'Religion and a Rights-Based Approach to Development', *Progress in Development Studies* 16 (2): 93–108.

Tomalin, Emma (2006b) 'The Thai Bhikkhuni Movement and Women's Empowerment', *Gender and Development* 14 (3): 385–397.

Tomalin, Emma (2007) *Sociology, Religion and Development: Literature Review*, Birmingham: Religions and Development WP 4.

Tomalin, Emma (2009a) 'Buddhist Feminist Transnational Networks, Female Ordination and Women's Empowerment', *Oxford Development Studies* 37 (2): 81–100.

Tomalin, Emma (2009b) *Hinduism and International Development: Religions and Development Background Paper*, Birmingham: Religions and Development WP 19.

Tomalin, Emma (2009c) *Bio-divinity and Biodiversity: The Limits to Religious Environmentalism*, Aldershot, UK: Ashgate.

Tomalin, Emma (2011) *Gender, Faith and Development*, Oxford: Oxfam; Rugby, UK: Practical Action Publishing.

Tomalin, Emma and Robert Leurs (2011) *Mapping the Development Activities of Local Faith-Based Organizations in Pakistan, India, Nigeria and Tanzania: A Comparative Analysis*, Birmingham: Religions and Development WP 62.

Torry, Malcolm (2005) *Managing God's Business: Religious and Faith-Based Organizations and Their Management*, Aldershot, UK: Ashgate.

Traer, Robert (1991) *Faith in Human Rights: Support in Religious Traditions for a Global Struggle*, Washington, DC: Georgetown University Press.

Tripp, Linda (1999) 'Gender and Development from a Christian Perspective: Experience from World Vision', *Gender and Development* 7 (1): 62–68.

Tucker, Mary Evelyn and Duncan Ryuken Williams (eds) (1997) *Buddhism and Ecology: The Interconnectedness of Dharma and Deeds*, Cambridge, MA: Harvard Divinity School, Center for the Study of World Religions/Harvard University Press.

Tylor, Edward B. (1958 [1871]) *Primitive Culture*, London: Murray.

Tyndale, Wendy (2006) *Visions of Development: Faith-Based Initiatives*. Aldershot, UK: Ashgate.

UNAIDS (2010) *Global Report: UNAIDS Report on the Global Aids Epidemic*. Online, available at: http://www.unaids.org/documents/20101123_GlobalReport_em.pdf (last accessed 11 March 2013).

UNDP (2006) *Applying a Human Rights-Based Approach to Development Cooperation and Programming: A UNDP Capacity Development Resource*. Online, available at: http://lencd.com/data/docs/252-Applying%20a%20Human%20Rights-based%20 20 Approach%20to%20Development%20Co.pdf (last accessed 13 September 2012).

UNDP (2011) *Fast Facts: Gender Equality and UNDP*. Online, available at: www.beta. undp.org/content/undp/en/home/librarypage/results/fast_facts/ff_gender_equality. html (last accessed 13 September 2012).

UNEP (2004) *Women and the Environment*, Nairobi: United Nations Environment Programme. Online, available at: www.unep.org/Documents.Multilingual/Default. asp?DocumentID=468&ArticleID=4488&l=en (last accessed 3 December 2012).

UNFPA (2005) *Culture Matters: Working with Communities and Faith-based Organizations: Case Studies from Country Programmes*, New York: UNFPA. Online, available at: www.unfpa.org/public/publications/pid/1430 (last accessed 13 September 2012).

UNFPA (2007) *Engaging Faith-Based Organizations in HIV Prevention: A Training Manual for Programme Managers*, New York: UNFPA. Online, available at: www. unfpa.org/public/global/pid/398 (last accessed 13 September 2012).

UNFPA (2008) *Culture Matters: Lessons from a Legacy of Engaging Faith-Based Organizations*, New York: UNFPA. Online, available at: www.unfpa.org/public/pid/ 1353 (last accessed 13 September 2012).

UNFPA (2009) *Guidelines for Engaging Faith-Based Organizations as Cultural Agents of Change*. Online, available at: http://www.unfpa.org/culture/docs/fbo_engagement. pdf (last accessed 12 March 2013).

United Nations (1987a) *Resolution adopted by the General Assembly* (from 42/187. Report of the World Commission on Environment and Development). Online, available at: http://www.un-documents.net/a42r187.htm (last accessed 11 March 2013).

United Nations (1987b) *Our Common Future, Chapter 2: Towards Sustainable Development* (from A/42/427. Our Common Future: Report of the World Commission on Environment and Development). Online, available at: http://www.un-documents.net/ocf-02.htm (last accessed 11 March 2013).

Vanderwoerd, Jim R. (2004) 'How Faith-Based Social Service Organizations Manage Secular Pressure Associated with Government Funding', *Nonprofit Management and Leadership* 14 (3): 239–262.

Ver Beek, Kurt (2000) 'Spirituality: A Development Taboo', *Development in Practice* 10 (1): 31–43.

Verhelst, Thierry and Wendy Tyndale (2002) 'Cultures, Spirituality and Development', in Deborah Eade (ed.) *Development and Culture*, Oxford: Oxfam GB in Association with World Faiths Development Dialogue, pp. 1–24.

Vuola, Elina (2002) 'Remaking Universals? Transnational Feminism(s) Challenging Fundamentalist Ecumenism', *Theory, Culture and Society* 19 (1–2): 175–195.

Waltz, Susan (2001) 'Universalizing Human Rights: The Role of Small States in the Construction of the Universal Declaration of Human Rights', *Human Rights Quarterly* 23 (1): 44–72.

Waltz, Susan (2002) 'Reclaiming and Rebuilding the History of the Universal Declaration of Human Rights', *Third World Quarterly* 23 (3): 437–448.

Warren, Karen (1987) 'Feminism and Ecology: Making Connections', *Environmental Ethics* 9 (3): 3–20.

Warren, Karen (1996) *Ecological Feminist Philosophies*, Bloomington: Indiana University Press.

WCED (United Nations World Commission on Environment and Development) (1987) *Our Common Future* (The Brundtland Report), Oxford: Oxford University Press.

Weber, Max (1930 [1904–1905]) *The Protestant Ethic and the Spirit of Capitalism*, London: Allen & Unwin.

Weber, Max (1958) *The Religions of India: The Sociology of Hinduism and Buddhism*, New York: Free Press.

Weber, Max (1963 [1922]) *The Sociology of Religion*, Boston: Beacon Press.

Whaling, Frank (1986) *Christian Theology and World Religions: A Global Approach*, Basingstoke, UK: Marshall Pickering.

White, Lynn Jr (1967) 'The Historical Roots of Our Environmental Crisis', *Science* 155 (3767): 1203–1207.

White, Ronald Cedric and Charles Howard Hopkins (1976) *The Social Gospel: Religion and Reform in Changing America*, Philadelphia: Temple University Press.

Whitten, Tony and Bryony Morgan (eds) (2006) *Faiths and the Environment*, Washington, DC: World Bank.

WHO (2006) *Appreciating Assets: The Contribution of Religion to Universal Access in Africa*. Online, available at: www.arhap.uct.ac.za/downloads/ARHAPWHO_entire. pdf (last accessed 13 September 2012).

Wilber, Charles and Kenneth Jameson (1980) 'Religious Values and the Social Limits to Development', *World Development* 8 (7/8): 467–479.

Wilson, Bryan (1985) 'Secularization: The Inherited Model', in Phillip E. Hammond (ed.) *The Sacred in the Secular Age: Toward Revision in the Scientific Study of Religion*, Berkeley: University of California Press, pp. 9–20.

Wilson, Rodney (1997) *Economics, Ethics and Religion: Jewish, Christian, and Muslim Economic Thought*, Basingstoke, UK: Macmillan; New York: New York University Press.

Witte, John Jr and Frank S. Alexander (eds) (2011) *Christianity and Human Rights: An Introduction*, Cambridge: Cambridge University Press.

Wood, R. L. (1999) 'Religious Culture and Political Action', *Sociological Theory* 17 (3): 307–332.

Woodhead, Linda (2004) *Christianity: A Very Short Introduction*, Oxford: Oxford Paperbacks.

Woodhead, Linda and Paul Heelas (eds) (2000) *Religion in Modern Times: An Interpretive Anthology*, Oxford: Blackwell.

World Bank (2006) *Faiths and the Environment: World Bank Support 2000–05*, Washington, DC: World Bank. Online, available at: www.arcworld.org/downloads/FaithsandEnvironment%20adobe.pdf (last accessed 3 January 2012).

World Bank (2011) *World Development Report 2012: Gender Equality and Development*, Washington, DC: World Bank. Online, available at: http://wdronline.worldbank.org/includes/imp_images/book_pdf/WDR_2012.pdf (last accessed 22 December 2011).

Wuthnow, Robert (2004) *Saving America? Faith-Based Services and the Future of Civil Society*, Princeton, NJ: Princeton University Press.

WWF (1986) *The Assisi Declarations: Messages on Man and Nature from Buddhism, Christianity, Hinduism, Islam and Judaism*, London: WWF.

Zadek, S. (1993) 'The Practice of Buddhist Economics?', *American Journal of Economics and Sociology* 52 (4): 433–445.

Zadek, S. and Szabo, S. (1993) 'Buddhist Organisation: The Case of the Sarvodaya Shramadana Movement', Value-Based Human Organisation Working Paper Series 2, London: New Economics Foundation.

Zaman, Asad (2008) *Islamic Economics: A Survey of the Literature*, Birmingham: Religions and Development WP 22.

Ziai, Aram (2004) 'The Ambivalence of Post-development: Between Reactionary Populism and Radical Democracy', *Third World Quarterly* 25 (6): 1045–1060.

Zinnbauer, Brian J., Kenneth I. Pargament, Brenda Cole, Mark S. Rye, Eric M. Butter, Timothy G. Belavich, Kathleen M. Hipp, Allie B. Scott and Jill L. Kadar (1997) 'Religion and Spirituality: Unfuzzying the Fuzzy', *Journal for the Scientific Study of Religion* 36 (4): 549–564.

Index